Conversion Factors

CONVERSION FACTORS

James L. Cook

Oxford New York Tokyo
OXFORD UNIVERSITY PRESS

Oxford University Press, Walton Street, Oxford OX2 6DP

Oxford New York Toronto
Delhi Bombay Calcutta Madras Karachi
Kuala Lumpur Singapore Hong Kong Tokyo
Nairobi Dar es Salaam Cape Town
Melbourne Auckland Madrid

and associated companies in
Berlin Ibadan

Oxford is a trade mark of Oxford University Press

Published in the United States by
Oxford University Press Inc., New York

First published 1991
Reprinted 1992, 1993

British Library Cataloguing in Publication Data
Cook, James L.
Conversion factors.
1. Technology. Conversion factors
I. Title
601.5
ISBN 0–19–856349–3
ISBN 0–19–856352–3 (pbk)

Library of Congress Cataloging in Publication Data
Cook, James L.
Conversion factors / James L. Cook.
Includes bibliographical references and index.
1. Metric system—Conversion tables.   2. Science—Tables.
3. Engineering—Tables.   I. Title
QC94.C67   1991   502.12—dc20   90-24853
ISBN 0–19–856349–3
ISBN 0–19–856352–3 (pbk)

Printed in Malta by Interprint Limited

To
Angela and Lynda

Introduction

This book should prove to be a handy reference for all students and teachers of physics, science, engineering, and other technical or academic disciplines, in addition to all those who work in industry.

More than *three thousand eight hundred* conversion factors have been prepared by the author for publication. About ninety per cent have been calculated from first principles, using primary conversion factors that can be found in most textbooks on the subject matter.

Primary conversion factors are identified with one or two asterisks for anyone who wishes to calculate their own unique relationships. These have been filtered out and compiled in a separate list on p. 105 for the convenience of the reader. Inexact numbers are marked with two asterisks (**), whilst exact numbers are given one asterisk (*).

The author has adopted two forms of notation for the display of all numbers. Proper fractions are used for better accuracy and because they sometimes avoid a recurring digit. Floating point notation expressed as a decimal fraction is used wherever practical, while scientific notation is used only when necessary. The number of digits is limited to ten for all conversion factors.

The relative accuracy of a number may be recognized by the notation with which it is displayed. Proper fractions are the most accurate. Numbers expressed as decimal fractions with less than ten digits may be rounded off to ten digits with zeros and therefore represent the second most accurate. Decimal fractions with all ten digits displayed are third in line of accuracy, followed by those displayed in scientific notation showing the mantissa ending with a zero. Finally, those numbers given in scientific notation that do not end with a zero should be used with some caution. However, the author has taken the step of including a few uncommon conversion factors merely to reveal hidden digits of other numbers which had to be truncated elsewhere in the same group. For example, see Statmhos to siemens units and Statmhos to picosiemens units. Therefore, greater accuracy can be found beyond the ten digits given in some cases.

Introduction

The information is grouped alphabetically into unit types to assist the specialist in finding any conversion factor with relative ease. In many cases, the reciprocal of a conversion factor has been printed elsewhere in the book to save the user the inconvenience of this calculation. It should therefore only be necessary for you to enter the *multiply* column with the unit that you wish to convert *from* and match it across the row in the *to obtain* column with the unit that you want to convert *to*. The conversion factor is found in the *by* column between the two units.

Although every effort has been made to calculate these factors with reliable consistency, neither the author nor the publisher shall be held responsible for their application or their use.

The author and publisher would be grateful to be notified of any corrections, which will be included in the next printing.

Fife   J.L.C.
March 1990

Contents

Contents

Contents

Abbreviations

Wherever possible, complete descriptive words are used in the tables for clarity but in some cases an abbreviation has been used to compress the data. A list of abbreviations is given below.

abs	absolute
accel'n	acceleration
A	ampere
apoth	apothecary
atm	standard atmosphere
atmos	atmospheric pressure
at	metric atmosphere
avoir	avoirdupois
Btu	British thermal unit
°C	Celsius (temperature)
Chu	Centigrade heat unit
cal	calorie
CGS or cgs	centimetre–gram–second
chrg	charge
cm	centimetre
cp	candle power
dia or diam	diameter
efi	electric field intensity
elec	electric
EM	electromotive
ES	electrostatic
evap	evaporated
f	force
°F	Fahrenheit (temperature)
ft	feet or foot
Hg	mercury
h	hour
hp	horsepower
Imp	Imperial
Int	international

Abbreviations

in	inch
IT	international tables
K	kelvin (temperature)
kg	kilogram (mass)
kgf	kilogram (force)
kilocal	kilocalories
kmol	kilomol
lb	pounds (weight)
liq	liquid
mfd	magnetic flux density
mfi	magnetic field intensity
min	minute (time)
MKS or mks	metre–kilogram–second
mmf	magnetomotive force
mN	millinewtons
nr	non-rationalized
psi	pounds/square inch
r	rationalized
R	Rankin (temperature)
RPM	revolutions per minute
s.c.d.	surface current density
s.ch.d.	surface charge density
s	second (time)
sq	square *or* squared
stp	Standard Temperature and Pressure
temp	temperature
thou	1 thousandth of an inch
v.c.d.	volume charge density
UK	United Kingdom
US	United States of America
yr	years
(15)	15.0 °C gram calorie

DIMENSIONS

M	mass
L	length
T	time
C	charge (electrical)

PREFIXES FOR (SI UNIT) MULTIPLES AND SUBMULTIPLES

a	atto	10^{-18}	da	deca	10^{1}
f	femto	10^{-15}	h	hecta	10^{2}
p	pico	10^{-12}	k	kilo	10^{3}
n	nano	10^{-9}	M	mega	10^{6}
μ	micro	10^{-6}	G	giga	10^{9}
m	milli	10^{-3}	T	tera	10^{12}
c	centi	10^{-2}	P	peta	10^{15}
d	deci	10^{-1}	E	exa	10^{18}

THE GREEK ALPHABET

A	α		alpha	= a		N	ν		nū	= n
B	β		bēta	= b		Ξ	ξ		xī	= x
Γ	γ		gamma	= g		O	o		omīcron	= o
Δ	δ		delta	= d		Π	π		pī	= p
E	ϵ		epsīlon	= e		P	ϱ		rhō	= r
Z	ζ		zēta	= z		Σ	σ	s	sigma	= s
H	η		ēta	= ē		T	τ		tau	= t
Θ	θ	ϑ	thēta	= th		Υ	υ		ūpsīlon	= u
I	ι		iōta	= i		Φ	ϕ		phī	= ph
K	$\varkappa$		kappa	= k		X	χ		chī	= kh
Λ	λ		lambda	= l		Ψ	ψ		psī	= ps
M	μ		mū	= m		Ω	ω		ōmega	= ō

Conversion notes

The four systems of units which are still in common use are:

1. British Imperial system
2. Centimetre–Gram–Second (CGS or cgs) system
3. Metre–Kilogram–Second (MKS or mks) system
4. Systeme International d'Unites (SI)

By far the most coherent of these is the SI system and although the author has included units from all the above systems in this book, he would strongly recommend that the reader learns to use the SI system at all times. This book should help in that respect.

Some definitions in this book differ slightly from those given in other references. The following notes have been compiled to clarify any differences that may concern the reader.

Note 1

The density of water is frequently quoted as 1000 kg/cubic metre or 1 gram/cubic centimetre with no reference to the conditions at which this density is valid. The calculation of an 'ideal' pressure column of water is given in BS 350, referring only to the 'standard gravity' of 9.80665 metres/second squared. For most practical purposes, water may be assumed incompressible but the temperature will have some effect on its density and for critical work should always be considered.

Conversion factors in this book which involve the density of water apply only at the temperature specified in the unit.

For example, a 1 millimetre column of water at 4.0 degrees Celsius and standard gravity may be calculated as follows:

1 mm of water = $0.0010 \times 999.940\,003\,6 \times 9.806\,65 = 9.806\,061\,636$ pascals

the units are = (metres) × (kg/cubic metre) × (metres/second²)
= (newtons/square metre) = pascals.

Conversion Notes

Note 2

The National Weights and Measures Laboratory (UK) quotes the UK gallon as exactly equal to 4.546 09 cubic decimetres (litres) as defined in Schedule 1, Part 4, of the Weights & Measures Act, 1985. Previously, the UK gallon had been defined as the space occupied by 10 pounds weight of distilled water of density 998.859 kg/cubic metre, weighed in air having a density of 1.217 kg/cubic metre and against weights of density 8136.0 kg/cubic metre. At these conditions, the temperature of the water may be calculated and it is found to be approximately 15.18 degrees Celsius. In this book, any reference to this temperature of water assumes the density of 998.859 kg/cubic metre.

Note 3

BS 350 defines the pound as exactly equal to 0.453 592 37 kilograms and 1 Btu/lb, exactly equal to 2.326 joules/gram. In this book, the pound is equal to 453.592 374 5 grams and the Btu_{IT} equivalent to 1055.055 863 joules.

Therefore: 1 Btu_{IT}/lb = 1055.055 863/453.592 374 5 joules/gram
$\qquad\qquad$ = 2.326 0 joules/gram, exactly to 9 decimal places.

Note 4

The section on *time* uses the following definitions:

the tropical year = 365 days, 5 hours, 48 minutes and 46 seconds
$\qquad\qquad\qquad$ = 365.242 199 074 days

the sidereal year = 365 days, 6 hours, 9 minutes and 9 seconds
$\qquad\qquad\qquad$ = 365.256 354 167 days.

Note 5

The author's adopted method of displaying numbers is considered to be more· flexible than simply limiting the number of decimal places after the point to four, five, or six.

The readers may truncate any number that requires a lesser degree of accuracy than that given. The floating point notation lends itself to better accuracy for reasons given in the Introduction.

Conversion factors

ACCELERATION (Angular)
Dimensions: $1/T^2$

MULTIPLY	BY	TO OBTAIN
radians/minute²	0.159 154 943	revolutions/minute²
radians/minute²	0.002 652 582	revolutions/min second
radians/minute²	4.4210×10^{-5}	revolutions/second²
radians/second²	572.957 795 2	revolutions/minute²
radians/second²	9.549 296 587	revolutions/min second
radians/second²	0.159 154 943	revolutions/second²
revolutions/minute²	0.001 745 329	radians/second²
revolutions/minute²	1/60	revolutions/min second
revolutions/minute²	1/3600	revolutions/second²
revolutions/second²	22 619.467 11	radians/minute²
revolutions/second²	6.283 185 307	radians/second²
revolutions/second²	3 600.0	revolutions/minute²
revolutions/second²	60.0	revolutions/min second

ACCELERATION (Linear)
Dimensions: L/T^2

MULTIPLY	BY		TO OBTAIN
accel'n by gravity,(g)	980.6650	†	centimetres/second²
accel'n by gravity,(g)	32.174 048 56	†	feet/second²
accel'n by gravity,(g)	* 9.806 650	†	metres/second²
centimetres/second²	0.032 808 399		feet/second²
centimetres/second²	0.0360		kilometres/hour second
centimetres/second²	0.010		metres/second²
centimetres/second²	0.022 369 363		miles/hour/second
feet/second²	30.480		centimetres/second²
feet/second²	1.097 280		kilometres/hour second
feet/second²	0.304 80		metres/second²
feet/second²	0.681 818 182		miles/hour second
kilometres/hour second	27.777 777 78		centimetres/second²

* denotes an exact number; ** denotes an inexact number. See the Introduction for further discussion.
† local gravity varies.

Acceleration (Linear)

MULTIPLY	BY	TO OBTAIN
kilometres/hour second	0.911 344 415	feet/second^2
kilometres/hour second	0.277 777 778	metres/second^2
kilometres/hour second	0.621 371 192	miles/hour second
metres/second^2	100.0	centimetres/second^2
metres/second^2	3.280 839 895	feet/second^2
metres/second^2	3.60	kilometres/hour second
metres/second^2	2.236 936 292	miles/hour second
miles/hour minute	0.745 066 667	centimetres/second^2
miles/hour second	44.7040	centimetres/second^2
miles/hour second	1.466 666 667	feet/second^2
miles/hour second	1.609 344 0	kilometres/hour second
miles/hour second	0.447 040	metres/second^2

AREA
Dimensions: L^2

MULTIPLY	BY	TO OBTAIN
acres	40.468 564 46	ares
acres	0.404 685 645	hectares
acres	* 4.0	roods (UK)
acres	4.046 86 × 10^7	square centimetres
acres	10.0	square chains (surveyors')
acres	43 560.0	square feet
acres	43 559.826	square feet (US Survey)
acres	6.272 64 × 10^6	square inches
acres	0.004 046 856	square kilometres
acres	100 000.0	square links (surveyors')
acres	4046.856 421	square metres
acres	0.001 562 50	square miles (statute)
acres	160.0	square perches
acres	160.0	square rods
acres	* 4840.0	square yards
ares	0.024 710 538	acres
ares	0.0100	hectares
ares	1.0 × 10^6	square centimetres
ares	1.0	square dekametres
ares	1076.391 042	square feet
ares	1076.386 736	square feet (US Survey)
ares	155 000.3100	square inches
ares	1.0 × 10^{-4}	square kilometres
ares	* 100.0	square metres

Area

MULTIPLY	BY	TO OBTAIN
ares	3.8610×10^{-5}	square miles
ares	119.599 0047	square yards
centares (centiares)	0.01	ares
centares (centiares)	10.763 910 42	square feet
centares (centiares)	1550.003 10	square inches
centares (centiares)	* 1.0	square metres
centares (centiares)	1.195 990 047	square yards
circular inches	645.160 000 1	circular millimetres
circular inches	1.0×10^6	circular mils
circular inches	5.067 074 792	square centimetres
circular inches	0.785 398 164	square inches
circular inches	$7.853\ 98 \times 10^5$	square mils
circular millimetres	0.007 853 982	square centimetres
circular millimetres	0.001 217 370	square inches
circular millimetres	0.785 398 163	square millimetres
circular mils	1.0×10^{-6}	circular inches
circular mils	5.0671×10^{-6}	square centimetres
circular mils	7.8540×10^{-7}	square inches
circular mils	0.000 506 707	square millimetres
circular mils	0.785 398 164	square mils
diameter(sphere) squared	3.141 592 654	surface area of sphere
hectares	2.471 053 816	acres
hectares	100.0	ares
hectares	1.0×10^8	square centimetres
hectares	107 639.1042	square feet
hectares	$1.550\ 00 \times 10^7$	square inches
hectares	0.010	square kilometres
hectares	10 000.0	square metres
hectares	0.003 861 022	square miles
hectares	395.368 610 5	square rods
hectares	11 959.900 47	square yards
major axis × minor axis	0.785 398 164	area of ellipse
roods (UK)	0.250	acres
roods (UK)	10.117 141 105	ares
roods (UK)	* 40.0	square perches
roods (UK)	1210.0	square yards
sections	1.0	square miles
square centimetres	2.4711×10^{-8}	acres
square centimetres	1.0×10^{-6}	ares
square centimetres	127.323 954 5	circular millimetres
square centimetres	197 352.5241	circular mils

Area

MULTIPLY	BY	TO OBTAIN
square centimetres	1.0×10^{-8}	hectares
square centimetres	1.0764×10^{-7}	square chains (engineers')
square centimetres	2.4711×10^{-7}	square chains (surveyors')
square centimetres	0.010	square decimetres
square centimetres	0.001 076 391	square feet
square centimetres	0.001 076 387	square feet (US Survey)
square centimetres	0.155 000 310	square inches
square centimetres	1.0×10^{-10}	square kilometres
square centimetres	0.000 10	square metres
square centimetres	3.861×10^{-11}	square miles
square centimetres	100.0	square millimetres
square centimetres	155 000.3100	square mils
square centimetres	3.9537×10^{-6}	square rods
square centimetres	0.000 119 599	square yards
square chains (engineers')	0.229 568 411	acres
square chains (engineers')	10 000.0	square feet
square chains (engineers')	9999.960	square feet (US Survey)
square chains (engineers')	1.4400×10^{6}	square inches
square chains (engineers')	10 000.0	square links (engineers')
square chains (engineers')	929.030 399 7	square metres
square chains (engineers')	0.000 358 701	square miles
square chains (engineers')	36.730 945 82	square rods
square chains (engineers')	1111.111 111	square yards
square chains (surveyors')	0.10	acres
square chains (surveyors')	4356.0	square feet
square chains (surveyors')	4355.982 576	square feet (US Survey)
square chains (surveyors')	627 264.0	square inches
square chains (surveyors')	10 000.0	square links (surveyors')
square chains (surveyors')	404.685 642 1	square metres
square chains (surveyors')	0.000 156 250	square miles
square chains (surveyors')	16.0	square rods
square chains (surveyors')	484.0	square yards
square decimetres	100.0	square centimetres
square decimetres	15.500 031 00	square inches
square dekametres	0.024 710 538	acres
square dekametres	1.0	ares
square dekametres	100.0	square metres
square dekametres	119.599 004 7	square yards
square feet	2.2957×10^{-5}	acres
square feet	0.000 929 030	ares
square feet	$1.833\ 46 \times 10^{8}$	circular mils

4

Area

MULTIPLY	BY	TO OBTAIN
square feet	9.2903×10^{-6}	hectares
square feet	929.030 399 7	square centimetres
square feet	0.000 229 568	square chains (surveyors')
square feet	0.999 996 0	square feet (US Survey)
square feet	144.0	square inches
square feet	9.2903×10^{-8}	square kilometres
square feet	2.295 684 114	square links (surveyors')
square feet	0.092 903 040	square metres
square feet	3.5870×10^{-8}	square miles
square feet	92 903.040	square millimetres
square feet	0.003 673 095	square rods
square feet	1/9	square yards
square feet (US Survey)	2.2957×10^{-5}	acres
square feet (US Survey)	22.956 932 97	microacres
square feet (US Survey)	929.034 116 1	square centimetres
square feet (US Survey)	0.000 100 000	square chains (engineers')
square feet (US Survey)	** 1.000 004 0	square feet
square hectometres	10 000.0	square metres
square inches	1.5942×10^{-7}	acres
square inches	6.4516×10^{-6}	ares
square inches	1.273 239 545	circular inches
square inches	$1.273\ 24 \times 10^{6}$	circular mils
square inches	6.4516×10^{-8}	hectares
square inches	6.451 60	square centimetres
square inches	1.5942×10^{-6}	square chains (surveyors')
square inches	0.064 516 0	square decimetres
square inches	0.006 944 444	square feet
square inches	0.006 944 417	square feet (US Survey)
square inches	6.452×10^{-10}	square kilometres
square inches	0.015 942 251	square links (surveyors')
square inches	0.000 645 160	square metres
square inches	2.491×10^{-10}	square miles
square inches	645.160	square millimetres
square inches	1.0×10^{6}	square mils
square inches	0.000 771 605	square yards
square kilometres	247.105 381 6	acres
square kilometres	10 000.0	ares
square kilometres	1.9736×10^{15}	circular mils
square kilometres	100.0	hectares
square kilometres	1.0×10^{10}	square centimetres
square kilometres	$1.076\ 39 \times 10^{7}$	square feet

Area

MULTIPLY	BY	TO OBTAIN
square kilometres	$1.076\ 39 \times 10^7$	square feet (US Survey)
square kilometres	1.550×10^9	square inches
square kilometres	1.0×10^6	square metres
square kilometres	0.386 102 159	square miles
square kilometres	1.0×10^{12}	square millimetres
square kilometres	$1.195\ 99 \times 10^6$	square yards
square links (engineers')	2.2957×10^{-5}	acres
square links (engineers')	929.0304	square centimetres
square links (engineers')	0.000 10	square chains (engineers')
square links (engineers')	1.0	square feet
square links (engineers')	0.999 996 0	square feet (US Survey)
square links (engineers')	144.0	square inches
square links (surveyors')	1.0×10^{-5}	acres
square links (surveyors')	404.685 642 1	square centimetres
square links (surveyors')	0.000 10	square chains (surveyors')
square links (surveyors')	0.435 60	square feet
square links (surveyors')	0.435 598 258	square feet (US Survey)
square links (surveyors')	62.726 40	square inches
square metres	0.000 247 105	acres
square metres	0.010	ares
square metres	1.0	centares
square metres	1.9735×10^9	circular mils
square metres	0.000 10	hectares
square metres	10 000.0	square centimetres
square metres	10.763 910 42	square feet
square metres	1550.003 100	square inches
square metres	1.0×10^6	square kilometres
square metres	10.763 910 42	square links (engineers')
square metres	24.710 538 15	square links (surveyors')
square metres	3.8610×10^{-7}	square miles
square metres	1.0×10^6	square millimetres
square metres	0.039 536 861	square rods
square metres	1.195 990 047	square yards
square miles	640.0	acres
square miles	25 899.881 09	ares
square miles	5.1114×10^{15}	circular mils
square miles	258.998 810 9	hectares
square miles	1.0	sections
square miles	2.5899×10^{10}	square centimetres
square miles	6400.0	square chains (surveyors')
square miles	$2.787\ 84 \times 10^7$	square feet

Area

MULTIPLY	BY	TO OBTAIN
square miles	$2.787\ 83 \times 10^7$	square feet (US Survey)
square miles	4.0145×10^9	square inches
square miles	2.589 988 110	square kilometres
square miles	$2.589\ 99 \times 10^6$	square metres
square miles	102 400.0	square rods
square miles	3.0976×10^6	square yards
square millimetres	1.273 239 545	circular millimetres
square millimetres	1973.525 241	circular mils
square millimetres	0.010	square centimetres
square millimetres	1.0764×10^{-5}	square feet
square millimetres	0.001 550 003	square inches
square millimetres	1.0×10^{-12}	square kilometres
square millimetres	1.0×10^{-6}	square metres
square millimetres	3.861×10^{-13}	square miles
square millimetres	1550.003 10	square mils
square millimetres	1.1956×10^{-6}	square yards
square mils	1.273 239 545	circular mils
square mils	6.4516×10^{-6}	square centimetres
square mils	6.9444×10^{-9}	square feet
square mils	1.0×10^{-6}	square inches
square mils	6.452×10^{-16}	square kilometres
square mils	6.452×10^{-10}	square metres
square mils	2.491×10^{-16}	square miles
square mils	0.000 645 160	square millimetres
square mils	7.716×10^{-10}	square yards
square rods	0.006 250	acres
square rods	0.252 928 526	ares
square rods	0.002 529 285	hectares
square rods	252 928.5263	square centimetres
square rods	272.250	square feet
square rods	272.248 911 0	square feet (US Survey)
square rods	39 204.0	square inches
square rods	272.250	square links (engineers')
square rods	625.0	square links (surveyors')
square rods	25.292 852 63	square metres
square rods	9.7656×10^{-6}	square miles
square rods	30.250	square yards
square yards	0.000 206 612	acres
square yards	0.008 361 274	ares
square yards	1.6501×10^9	circular mils
square yards	8.3613×10^{-5}	hectares

7

Area

MULTIPLY	BY	TO OBTAIN
square yards	8361.273 60	square centimetres
square yards	0.000 90	square chains (engineers')
square yards	0.002 066 116	square chains (surveyors')
square yards	9.0	square feet
square yards	8.999 964 0	square feet (US Survey)
square yards	1296.0	square inches
square yards	8.3613×10^{-7}	square kilometres
square yards	9.0	square links (engineers')
square yards	20.661 157 02	square links (surveyors')
square yards	0.836 127 360	square metres
square yards	3.2283×10^{-7}	square miles
square yards	$8.361\ 27 \times 10^5$	square millimetres
square yards	0.033 057 851	square perches
square yards	0.033 057 851	square rods

AREA FLOW
Dimensions: L^2/T

MULTIPLY	BY	TO OBTAIN
square centimetres/hour	1.7940×10^{-5}	square feet/minute
square centimetres/min	0.064 583 463	square feet/hour
square feet/hour	0.258 064 0	square centimetres/s
square feet/hour	2.5806×10^{-5}	square metres/second
square feet/minute	15.483 840	square centimetres/s
square feet/minute	0.001 548 384	square metres/second
square inches/second	23 225.760	sq centimetres/hour
square inches/second	6.451 60	sq centimetres/second
square inches/second	0.416 666 667	square feet/minute
square metres/hour	0.179 398 507	square feet/minute
square metres/minute	645.834 625 2	square feet/hour

CALORIFIC VALUE (Volume basis)
Dimensions: M/LT^2

MULTIPLY	BY	TO OBTAIN
Btu/cubic foot	37 258.946 17	joules/cubic metre
Btu/cubic foot	8.899 146 406	kilocal/cubic metre
Btu/cubic foot	37.258 946 17	kilojoules/cubic metre
Btu/cubic foot	0.037 258 946	megajoules/cubic metre
calories/cubic cm	$4.186\ 80 \times 10^6$	joules/cubic metre
calories/cubic cm	4186.80	kilojoules/cubic metre
calories/cubic cm	4.186 80	megajoules/cubic metre

Calorific Value (Volume basis)

MULTIPLY	BY	TO OBTAIN
Chu/cubic foot	67 066.103 10	joules/cubic metre
Chu/cubic foot	67.066 103 10	kilojoules/cubic metre
Chu/cubic foot	0.067 066 103	megajoules/cubic metre
kilocals/cubic metre	0.112 370 328	Btu/cubic foot
kilocals/cubic metre	4186.80	joules/cubic metre
kilocals/cubic metre	4.186 80	kilojoules/cubic metre
kilocals/cubic metre	0.004 186 80	megajoules/cubic metre
megajoules/cubic metre	1.0×10^6	joules/cubic metre
therms/cubic foot	$3.725\ 89 \times 10^9$	joules/cubic metre
therms/cubic foot	$3.725\ 89 \times 10^6$	kilojoules/cubic metre
therms/cubic foot	3725.894 617	megajoules/cubic metre
therms/gallon (UK)	2.3208×10^{10}	joules/cubic metre

CALORIFIC VALUE (Volume, British Gas)
Dimensions: M/LT^2

MULTIPLY	BY	TO OBTAIN
Btu(15)/cubic foot	8.899 146 410	kg cal(15)/cubic metre
Btu(15)/cubic foot	0.037 247 377	megajoules/cubic metre
Btu(15)/cubic foot	0.0010	therms/100 cubic feet
Btu(15)/cubic foot	1.0×10^{-5}	therms/cubic foot

CHEMICAL (Henry's law)
Dimensions: L^2/T^2

MULTIPLY	BY	TO OBTAIN
atm/(grams/cubic cm)	0.001 013 250	bars/(kg/cubic metre)
atm/(grams/cubic cm)	101.3250	newton metres/kilogram
atm/(kg/cubic foot)	0.028 692 045	bars/(kg/cubic metre)
atm/(kg/cubic foot)	2869.204 481	newton metres/kilogram
atm/(kg/cubic metre)	1.013 250	bars/(kg/cubic metre)
atm/(kg/cubic metre)	101 325.0	newton metres/kilogram
atm/(lb/cubic foot)	0.063 255 130	bars/(kg/cubic metre)
atm/(lb/cubic foot)	6325.513 043	newton metres/kilogram
bars/(kg/cubic metre)	100 000.0	newton metres/kilogram
newton metres/kilogram	0.000 010	bars/(kg/cubic metre)

CIRCULAR and SPHERICAL GEOMETRY
Dimensions: Various

MULTIPLY	BY	TO OBTAIN
circles	1.0	circumferences
circles	360.0	degrees

Circular and Spherical Geometry

MULTIPLY	BY	TO OBTAIN
circles	400.0	grades
circles	21 600.0	minutes
circles	6.283 185 308	radians
circles	* 1.0	revolutions
circles	1.2960×10^6	seconds
circumferences	1.0	circles
circumferences	360.0	degrees
circumferences	400.0	grades
circumferences	21 600.0	minutes
circumferences	6.283 185 308	radians
circumferences	* 1.0	revolutions
circumferences	1.2960×10^6	seconds
degrees	1/360	circles
degrees	* 60.0	minutes
degrees	1/90	quadrants
degrees	0.017 453 293	radians
degrees	3600.0	seconds
diameter of circle	3.141 592 654	circumference
diameter of circle	0.707 106 781	square side (inscribed)
diameter of circle	0.886 226 926	square side (equal area)
grades	0.002 50	circles
grades	0.002 50	circumferences
grades	0.90	degrees
grades	54.0	minutes
grades	0.015 707 963	radians
grades	0.002 50	revolutions
grades	3240.0	seconds
hexagon across corners	0.866 025 404	hexagon across flats
hexagon across flats	1.154 700, 538	hexagon across corners
minutes	1/60	degrees
minutes	1/5400	quadrants
minutes	0.000 290 888	radians
minutes	* 60.0	seconds
quadrants	* 90.0	degrees
quadrants	5400.0	minutes
quadrants	1.570 796 327	radians
quadrants	3.240×10^5	seconds
radians	0.159 154 943	circumferences
radians	** 57.295 779 51	degrees
radians	3437.746 771	minutes
radians	0.636 619 772	quadrants

Circular and Spherical Geometry

MULTIPLY	BY	TO OBTAIN
radians	0.159 154 943	revolutions
radians	206 264.8062	seconds
radians/centimetre	57.295 779 51	degrees/centimetre
radians/centimetre	1746.375 360	degrees/foot
radians/centimetre	145.531 280	degrees/inch
radians/centimetre	3437.746 771	minutes/centimetre
revolutions	1.0	circles
revolutions	1.0	circumferences
revolutions	* 360.0	degrees
revolutions	* 400.0	grades
revolutions	* 4.0	quadrants
revolutions	6.283 185 308	radians
seconds	1/3600	degrees
seconds	1/60	minutes
seconds	3.0864×10^{-6}	quadrants
seconds	4.8481×10^{-6}	radians
solid angles	12.566 370 62	steradians
spheres	2.0	hemispheres
spheres	* 1.0	solid angles
spheres	* 8.0	spherical right angles
spheres	** 12.566 370 62	steradians
spherical right angles	0.250	hemispheres
spherical right angles	0.1250	spheres
spherical right angles	1.570 796 328	steradians
square across corners	0.707 106 781	square across flats
square across flats	1.414 213 562	square across corners
steradians	0.159 154 943	hemispheres
steradians	0.079 577 472	solid angles
steradians	0.079 577 472	spheres
steradians	0.636 619 772	spherical right angles
steradians	3282.806 563	square degrees

COEFFICIENT OF EXPANSION (Volumetric)
Dimensions: M/L^3

MULTIPLY	BY	TO OBTAIN
grams/cubic cm °C	1000.0	kg/cubic metre °C
grams/cubic cm °C	1000.0	kg/cubic metre K
grams/cubic cm °C	62.427 959 95	pounds/cubic foot °C
grams/cubic cm °C	34.682 199 97	pounds/cubic foot °F
kg/cubic metre °C	0.0010	grams/cubic cm °C

11

Coefficient of Expansion (Volumetric)

MULTIPLY	BY	TO OBTAIN
kg/cubic metre °C	1.0	kg/cubic metre K
kg/cubic metre °C	0.062 427 960	pounds/cubic foot °C
kg/cubic metre °C	0.034 682 20	pounds/cubic foot °F
pounds/cubic foot °C	0.016 018 464	grams/cubic cm °C
pounds/cubic foot °C	16.018 463 53	kg/cubic metre °C
pounds/cubic foot °C	16.018 463 53	kg/cubic metre K
pounds/cubic foot °F	0.028 833 234	grams/cubic cm °C
pounds/cubic foot °F	28.833 234 35	kg/cubic metre °C
pounds/cubic foot °F	28.833 234 35	kg/cubic metre K

CONCENTRATION (Liquid solution)
Dimensions: M/L^3

MULTIPLY	BY	TO OBTAIN
grains/gallon (UK)	142.857 142 9	lbs/million UK gallons
grains/gallon (UK)	14.270 049 79	parts/million (ppm)
grains/gallon (US)	142.857 142 9	lbs/million US gallons
grains/gallon (US)	17.137 615 23	parts/million (ppm)
grams/litre	70.156 889 29	grains/gallon (UK)
grams/litre	58.417 830 60	grains/gallon (US)
grams/litre	1001.142 303	parts/million (ppm)
grams/litre	0.062 427 960	pounds/cubic foot
grams/litre	0.010 022 413	pounds/gallon (UK)
grams/litre	0.008 345 404	pounds/gallon (US)
milligrams/litre	0.058 417 831	grains/gallon (US)
milligrams/litre	0.0010	grams/litre
milligrams/litre	1.001 142 303	parts/million (ppm)
parts/million (ppm)	0.070 076 840	grains/gallon (UK)
parts/million (ppm)	0.058 351 176	grains/gallon (US)
parts/million (ppm)	0.000 998 859	grams/litre
parts/million (ppm)	10.010 977 18	lbs/million UK gallons
parts/million (ppm)	8.335 882 263	lbs/million US gallons
parts/million (ppm)	0.998 859 0	milligrams/litre

DENSITY (Mass concentration)
Dimensions: M/L^3

MULTIPLY	BY	TO OBTAIN
grains/cubic foot	2.288 351 934	grams/cubic metre
grains/cubic foot	0.002 288 352	kilograms/cubic metre
grams/cubic centimetre	980.6650	dynes/cubic centimetre
grams/cubic centimetre	436 995.7197	grains/cubic foot

Density (Mass concentration)

MULTIPLY	BY	TO OBTAIN
grams/cubic centimetre	15.432 358 20	grains/millilitre
grams/cubic centimetre	1.0	grams/millilitre
grams/cubic centimetre	1.0	kilograms/litre
grams/cubic centimetre	1000.0	kilograms/metre cubed
grams/cubic centimetre	1.162 361 236	poundals/cubic inch
grams/cubic centimetre	3.4049×10^{-7}	pounds/circular mil ft
grams/cubic centimetre	62.427 959 95	pounds/cubic foot
grams/cubic centimetre	0.036 127 292	pounds/cubic inch
grams/cubic centimetre	10.022 412 76	pounds/gallon (UK)
grams/cubic centimetre	9.711 106 311	pounds/gallon (US dry)
grams/cubic centimetre	8.345 404 375	pounds/gallon (US liq)
grams/cubic metre	0.436 995 720	grains/cubic foot
grams/cubic metre	0.0010	kilograms/cubic metre
grams/cubic metre	6.2428×10^{-5}	pounds/cubic foot
kilograms/cubic metre	0.0010	grams/cubic centimetre
kilograms/cubic metre	3.405×10^{-10}	pounds/circular mil ft
kilograms/cubic metre	0.062 427 960	pounds/cubic foot
kilograms/cubic metre	3.6127×10^{-5}	pounds/cubic inch
kilograms/cubic metre	0.010 022 413	pounds/gallon (UK)
kilograms/cubic metre	0.008 353 404	pounds/gallon (US)
kilograms/litre	1000.0	kilograms/cubic metre
milligrams/litre	0.0010	kilograms/cubic metre
milligrams/litre	6.2428×10^{-5}	pounds/cubic foot
pounds/circular mil ft	$2.936\ 93 \times 10^{6}$	grams/cubic centimetre
pounds/cubic foot	0.016 018 464	grams/cubic centimetre
pounds/cubic foot	5.4542×10^{-9}	pounds/circular mil ft
pounds/cubic foot	0.000 578 704	pounds/cubic inch
pounds/cubic ft (gas)	16.018 463 53	kilograms/cubic metre
pounds/cubic ft (liquid)	0.016 018 464	kilograms/litre
pounds/cubic ft (solid)	16.018 463 53	kilograms/cubic metre
pounds/cubic inch	27.679 904 98	grams/cubic centimetre
pounds/cubic inch	27 679.904 98	grams/litre
pounds/cubic inch	27 679.904 98	kilograms/cubic metre
pounds/cubic inch	9.4248×10^{-6}	pounds/circular mil ft
pounds/cubic inch	1728.0	pounds/cubic foot
pounds/gallon (UK)	0.099 776 374	kilograms/litre
pounds/gallon (UK)	99.776 373 65	kilograms/metre cubed
pounds/gallon (UK)	6.228 835 459	pounds/cubic foot
pounds/gallon (US liq)	0.119 826 429	grams/cubic centimetre
pounds/gallon (US liq)	119.826 428 5	kilograms/cubic metre
pounds/gallon (US liq)	0.119 826 429	kilograms/litre

Density (Mass concentration)

MULTIPLY	BY	TO OBTAIN
pounds/gallon (US liq)	7.480 519 479	pounds/cubic foot
slugs/cubic foot	0.515 378 823	grams/cubic centimetre

ELECTRICAL (Charge)
Dimensions: C

MULTIPLY	BY	TO OBTAIN
Abcoulombs	1/360	ampere hours
Abcoulombs	10.0	coulombs
Abcoulombs	6.2419×10^{19}	electronic charge
ampere hours	360.0	Abcoulombs
ampere hours	3600.0	coulombs
CGS e.m. units	1.0	Abcoulombs
CGS e.m. units	* 10.0	coulombs
CGS e.m. units	2.9979×10^{10}	Statcoulombs
CGS e.s. units	3.336×10^{-11}	Abcoulombs
CGS e.s. units	3.336×10^{-10}	coulombs
CGS e.s. units	* 1.0	Statcoulombs
coulombs	0.1	Abcoulombs
coulombs	1/3600	ampere hours
coulombs	* 1.0	ampere seconds
coulombs	6.2414×10^{18}	electronic charge
coulombs	$2.997 \, 92 \times 10^{9}$	Statcoulombs
coulombs (International)	0.999 835 027	coulombs
electronic charge	160.220×10^{-22}	Abcoulombs
electronic charge	160.220×10^{-21}	coulombs
electronic charge	480.326×10^{-12}	Statcoulombs
faradays (chem)	9648.998 385	Abcoulombs
faradays (chem)	** 26.802 773 29	ampere hours
faradays (chem)	96 489.983 84	coulombs
faradays (chem)	96 505.904 71	coulombs (International)
faradays (phys)	9651.708 450	Abcoulombs
faradays (phys)	** 26.810 301 25	ampere hours
faradays (phys)	96 517.084 50	coulombs
faradays (phys)	96 533.009 81	coulombs (International)
MKS units	1.0	coulombs
Statcoulombs	3.336×10^{-11}	Abcoulombs
Statcoulombs	9.266×10^{-14}	ampere hours
Statcoulombs	3.336×10^{-10}	coulombs
Statcoulombs	$2.081 \, 92 \times 10^{9}$	electronic charge
Statcoulombs	** 333.564 604 8	picocoulombs

14

ELECTRICAL (Conductance)
Dimensions: C^2T/ML^2

MULTIPLY		BY	TO OBTAIN
Abmhos		1000.0	megasiemens units
Abmhos		1.0×10^9	mhos
Abmhos		1.0×10^9	siemens
CGS e.m. units		1.0	Abmhos
CGS e.m. units	*	1.0×10^9	siemens
CGS e.s. units		1.113×10^{-21}	Abmhos
CGS e.s. units		1.113×10^{-12}	siemens
CGS e.s. units	*	1.0	Statmhos
mhos		1.0×10^{-6}	megasiemens units
mhos	*	1.0	siemens
mhos (International)		9.995×10^{-10}	Abmhos
mhos (International)		0.999 505 245	mhos
mhos (International)	**	0.999 505 245	siemens
MKS units		1.0	siemens
Statmhos		1.113×10^{-12}	mhos
Statmhos	**	1.112 653 456	picosiemens units
Statmhos		1.113×10^{-12}	siemens

ELECTRICAL (Conductivity)
Dimensions: C^2T/ML^3

MULTIPLY		BY	TO OBTAIN
Abmhos/centimetre		1.0×10^7	siemens/metre
Abmhos/centimetre		8.9875×10^{22}	Statmhos/metre
Abmhos/metre		1.0×10^9	siemens/metre
Abmhos/metre		8.9875×10^{18}	Statmhos/centimetre
CGS e.m. units		100.0	Abmhos/metre
CGS e.m. units	*	1.0×10^{11}	siemens/metre
CGS e.s. units		1.113×10^{-19}	Abmhos/metre
CGS e.s. units		1.113×10^{-10}	siemens/metre
CGS e.s. units	*	1.0	Statmhos/centimetre
mhos/centimetre		0.010	siemens/metre
mhos/metre		1.0	siemens/metre
mhos/metre		$8.987\ 52 \times 10^9$	Statmhos/centimetre
siemens/metre		1.0×10^{-11}	Abmhos/centimetre
siemens/metre		1.0×10^{-9}	Abmhos/metre
siemens/metre		0.010	mhos/centimetre
siemens/metre		$8.987\ 52 \times 10^9$	Statmhos/centimetre

Electrical (Conductivity)

MULTIPLY	BY	TO OBTAIN
Statmhos/centimetre	** 111.265 345 6	picosiemens/metre
Statmhos/centimetre	1.113×10^{-10}	siemens/metre

ELECTRICAL (Current)
Dimensions: C/T

MULTIPLY	BY	TO OBTAIN
Abamperes	* 10.0	amperes
Abamperes	* 10.0	coulombs/second
Abamperes	2.9979×10^{10}	Statamperes
amperes	0.1	Abamperes
amperes	2.9979×10^{9}	Statamperes
amperes (International)	** 0.999 835 027	amperes
amperes (International)	1.0	coulombs(Int)/second
CGS e.m. units	1.0	Abamperes
CGS e.m. units	* 10.0	amperes
CGS e.m. units	2.9979×10^{10}	Statamperes
CGS e.s. units	3.336×10^{-11}	Abamperes
CGS e.s. units	3.336×10^{-10}	amperes
CGS e.s. units	* 1.0	Statamperes
coulombs(Int)/second	0.999 835 027	amperes
coulombs/second	* 1.0	amperes
faradays(chem)/second	** 9648.998 385	Abamperes
faradays(chem)/second	96 489.983 85	amperes
faradays(chem)/second	96 505.904 71	amperes (International)
faradays(phys)/second	** 9651.708 450	Abamperes
faradays(phys)/second	96 517.084 50	amperes
faradays(phys)/second	96 533.009 82	amperes (International)
MKS units	1.0	amperes
Statamperes	3.336×10^{-11}	Abamperes
Statamperes	3.336×10^{-10}	amperes
Statamperes	** 333.564 604 8	picoamperes

ELECTRICAL (Electric capacitance)
Dimensions: C^2T^2/ML^2

MULTIPLY	BY	TO OBTAIN
Abfarads	* 1.0×10^{9}	farads
Abfarads	1.0×10^{15}	microfarads
Abfarads	8.9875×10^{20}	Statfarads
CGS e.m. units	1.0	Abfarads

Electrical (Electric capacitance)

MULTIPLY	BY	TO OBTAIN
CGS e.m. units	* 1.0×10^9	farads
CGS e.m. units	8.9875×10^{20}	Statfarads
CGS e.s. units	1.113×10^{-21}	Abfarads
CGS e.s. units	1.113×10^{-12}	farads
CGS e.s. units	* 1.0	Statfarads
farads	1.0×10^{-9}	Abfarads
farads	1.0×10^6	microfarads
farads	8.9875×10^{11}	Statfarads
farads (International)	** $0.999\ 505\ 245$	farads
microfarads	1.0×10^{-15}	Abfarads
microfarads	1.0×10^{-6}	farads
microfarads	8.9875×10^5	Statfarads
Statfarads	1.113×10^{-21}	Abfarads
Statfarads	1.113×10^{-12}	farads
Statfarads	1.1126×10^{-6}	microfarads
Statfarads	** $1.112\ 653\ 456$	picofarads

ELECTRICAL (Electric field)
Dimensions: ML/CT^2

MULTIPLY	BY	TO OBTAIN
Abvolts/centimetre	1.0×10^{-8}	volts/centimetre
Abvolts/centimetre	2.54×10^{-8}	volts/inch
Abvolts/centimetre	1.0×10^{-6}	volts/metre
CGS e.m. units	1.0	Abvolts/centimetre
CGS e.m. units	* 1.0×10^{-6}	volts/metre
CGS e.s. units	2.9979×10^{10}	Abvolts/centimetre
CGS e.s. units	$0.033\ 356\ 460$	millivolts/metre
CGS e.s. units	* 1.0	Statvolts/centimetre
CGS e.s. units	$29\ 979.20$	volts/metre
kilovolts/centimetre	1.0×10^{11}	Abvolts/centimetre
kilovolts/centimetre	1.0×10^{11}	microvolts/metre
kilovolts/centimetre	1.0×10^8	millivolts/metre
kilovolts/centimetre	$3.335\ 646\ 048$	Statvolts/centimetre
kilovolts/centimetre	2540.0	volts/inch
kilovolts/centimetre	$100\ 000.0$	volts/metre
Statvolts/centimetre	299.7920	volts/centimetre
Statvolts/centimetre	$761.471\ 680\ 0$	volts/inch
Statvolts/centimetre	$29\ 979.20$	volts/metre
Statvolts/inch	$118.028\ 346\ 5$	volts/centimetre
volts/inch	$0.393\ 700\ 787$	volts/centimetre

ELECTRICAL (Electric potential)
Dimensions: ML^2/CT^2

MULTIPLY	BY	TO OBTAIN
Abvolts	0.0100	microvolts
Abvolts	1.0×10^{-5}	millivolts
Abvolts	* 1.0×10^{-8}	volts
Abvolts	9.9967×10^{-9}	volts (International)
CGS e.m. units	1.0	Abvolts
CGS e.m. units	* 1.0×10^{-8}	volts
CGS e.s. units	2.9979×10^{10}	Abvolts
CGS e.s. units	* 1.0	Statvolts
CGS e.s. units	** 299.7920	volts
millivolts	3.3356×10^{-6}	Statvolts
millivolts	0.0010	volts
MKS units	1.0	volts
Statvolts	2.9979×10^{10}	Abvolts
Statvolts	299 792.0	millivolts
volts	1.0×10^{8}	Abvolts
volts	1.0	joules/coulomb
volts	0.003 335 635	Statvolts
volts	0.999 669 110	volts (International)
volts (International)	1.593×10^{-19}	joules/electron
volts (International)	** 1.000 331 0	volts
volts/metre	0.000 033 356	Statvolts/centimetre

ELECTRICAL (Inductance)
Dimensions: ML^2/C^2

MULTIPLY	BY	TO OBTAIN
Abhenries	* 1.0×10^{-9}	henries
CGS e.m. units	1.0	Abhenries
CGS e.m. units	* 1.0×10^{-9}	henries
CGS e.m. units	1.113×10^{-21}	Stathenries
CGS e.s. units	8.9876×10^{20}	Abhenries
CGS e.s. units	8.9876×10^{11}	henries
CGS e.s. units	* 1.0	Stathenries
henries	1.0×10^{9}	Abhenries
henries	1.0×10^{-9}	gigahenries
henries	0.999 505 245	henries (International)
henries	1000.0	millihenries
henries	1.113×10^{-12}	Stathenries
microhenries	1.0×10^{-6}	henries

Electrical (Inductance)

MULTIPLY	BY	TO OBTAIN
microhenries	1.113×10^{-18}	Stathenries
millihenries	1.0×10^6	Abhenries
millihenries	0.0010	henries
millihenries	1.113×10^{-15}	Stathenries
MKS(r or nr) units	1.0	henries
Stathenries	8.9875×10^{20}	Abhenries
Stathenries	** 898.752 432 4	gigahenries
Stathenries	8.9875×10^{11}	henries
Stathenries	898 752.4324	megahenries
Stathenries	8.9875×10^{14}	millihenries

ELECTRICAL (Linear charge density)
Dimensions: C/L

MULTIPLY	BY	TO OBTAIN
Abcoulombs/centimetre	2.540	Abcoulombs/inch
Abcoulombs/centimetre	100.0	Abcoulombs/metre
Abcoulombs/centimetre	10.0	coulombs/centimetre
Abcoulombs/centimetre	25.40	coulombs/inch
Abcoulombs/centimetre	1000.0	coulombs/metre
Abcoulombs/inch	0.393 700 787	Abcoulombs/centimetre
Abcoulombs/inch	39.370 078 74	Abcoulombs/metre
Abcoulombs/inch	3.937 007 874	coulombs/centimetre
Abcoulombs/inch	10.0	coulombs/inch
Abcoulombs/inch	393.700 787 4	coulombs/metre
Abcoulombs/metre	0.010	Abcoulombs/centimetre
Abcoulombs/metre	0.025 40	Abcoulombs/inch
Abcoulombs/metre	0.10	coulombs/centimetre
Abcoulombs/metre	0.2540	coulombs/inch
Abcoulombs/metre	10.0	coulombs/metre
coulombs/centimetre	0.10	Abcoulombs/centimetre
coulombs/centimetre	0.2540	Abcoulombs/inch
coulombs/centimetre	10.0	Abcoulombs/metre
coulombs/centimetre	2.540	coulombs/inch
coulombs/centimetre	100.0	coulombs/metre
coulombs/inch	0.039 370 079	Abcoulombs/centimetre
coulombs/inch	0.10	Abcoulombs/inch
coulombs/inch	3.937 007 874	Abcoulombs/metre
coulombs/inch	0.393 700 787	coulombs/centimetre
coulombs/inch	39.370 078 74	coulombs/metre
coulombs/metre	0.0010	Abcoulombs/centimetre

Electrical (Linear charge density)

MULTIPLY	BY	TO OBTAIN
coulombs/metre	0.002 540	Abcoulombs/inch
coulombs/metre	0.10	Abcoulombs/metre
coulombs/metre	0.010	coulombs/centimetre
coulombs/metre	0.025 40	coulombs/inch

ELECTRICAL (Linear current density)
Dimensions: C/LT

MULTIPLY	BY	TO OBTAIN
Abamperes/centimetre	2.540	Abamperes/inch
Abamperes/centimetre	100.0	Abamperes/metre
Abamperes/centimetre	10.0	amperes/centimetre
Abamperes/centimetre	25.40	amperes/inch
Abamperes/centimetre	1000.0	amperes/metre
Abamperes/inch	0.393 700 787	Abamperes/centimetre
Abamperes/inch	39.370 078 74	Abamperes/metre
Abamperes/inch	3.937 007 874	amperes/centimetre
Abamperes/inch	10.0	amperes/inch
Abamperes/inch	393.700 787 4	amperes/metre
Abamperes/metre	0.010	Abamperes/centimetre
Abamperes/metre	0.025 40	Abamperes/inch
Abamperes/metre	0.10	amperes/centimetre
Abamperes/metre	0.2540	amperes/inch
Abamperes/metre	10.0	amperes/metre
amperes/centimetre	0.10	Abamperes/centimetre
amperes/centimetre	0.2540	Abamperes/inch
amperes/centimetre	10.0	Abamperes/metre
amperes/centimetre	2.540	amperes/inch
amperes/centimetre	100.0	amperes/metre
amperes/inch	0.039 370 079	Abamperes/centimetre
amperes/inch	0.10	Abamperes/inch
amperes/inch	3.937 007 874	Abamperes/metre
amperes/inch	0.393 700 787	amperes/centimetre
amperes/inch	39.370 078 74	amperes/metre
amperes/metre	0.0010	Abamperes/centimetre
amperes/metre	0.002 540	Abamperes/inch
amperes/metre	0.10	Abamperes/metre
amperes/metre	0.010	amperes/centimetre
amperes/metre	0.025 40	amperes/inch
CSG e.m. units	79.577 471 51	amperes/metre
CGS e.m. units	* 1.0	oersteds

Electrical (Linear current density)

MULTIPLY	BY	TO OBTAIN
CGS e.s. units	2.6544×10^{-9}	amperes/metre
CGS e.s. units	3.335×10^{-11}	oersteds
CGS e.s. units	** 2 654.422 783	picoamperes/metre
gilberts/centimetre	1.0	oersteds
MKS units	1.0	amperes/metre
oersteds	0.795 774 715	amperes/centimetre
oersteds	2.021 267 776	amperes/inch
oersteds	79.577 471 51	amperes/metre
oersteds (International)	0.999 835 027	oersteds

ELECTRICAL (Magnetic field strength)
Dimensions: C/LT

MULTIPLY	BY	TO OBTAIN
Abamperes/centimetre	2.540	Abamperes/inch
Abamperes/centimetre	100.0	Abamperes/metre
Abamperes/centimetre	10.0	amperes/centimetre
Abamperes/centimetre	25.40	amperes/inch
Abamperes/centimetre	1000.0	amperes/metre
Abamperes/inch	0.393 700 787	Abamperes/centimetre
Abamperes/inch	39.370 078 74	Abamperes/metre
Abamperes/inch	3.937 007 874	amperes/centimetre
Abamperes/inch	10.0	amperes/inch
Abamperes/inch	393.700 787 4	amperes/metre
Abamperes/metre	0.010	Abamperes/centimetre
Abamperes/metre	0.025 40	Abamperes/inch
Abamperes/metre	0.10	amperes/centimetre
Abamperes/metre	0.2540	amperes/inch
Abamperes/metre	10.0	amperes/metre
amperes/centimetre	0.10	Abamperes/centimetre
amperes/centimetre	0.2540	Abamperes/inch
amperes/centimetre	10.0	Abamperes/metre
amperes/centimetre	2.540	amperes/inch
amperes/centimetre	100.0	amperes/metre
amperes/inch	0.039 370 079	Abamperes/centimetre
amperes/inch	0.10	Abamperes/inch
amperes/inch	3.937 007 874	Abamperes/metre
amperes/inch	0.393 700 784	amperes/centimetre
amperes/inch	39.370 078 40	amperes/metre
amperes/metre	0.0010	Abamperes/centimetre
amperes/metre	0.002 540	Abamperes/inch

Electrical (Magnetic field strength)

MULTIPLY	BY		TO OBTAIN
amperes/metre		0.10	Abamperes/metre
amperes/metre		0.010	amperes/centimetre
amperes/metre		0.025 40	amperes/inch
CGS e.m. units	**	79.577 471 51	amperes/metre
CGS e.m. units	*	1.0	oersteds
CGS e.s. units		2.6544×10^{-9}	amperes/metre
CGS e.s. units		3.335×10^{-11}	oersteds
CGS e.s. units	**	2654.422 783	picoamperes/metre
gilberts/centimetre		1.0	oersteds
MKS units		1.0	amperes/metre
oersteds		0.795 774 715	amperes/centimetre
oersteds		2.021 267 776	amperes/inch
oersteds		79.577 471 51	amperes/metre
oersteds (International)	**	0.999 835 027	oersteds

ELECTRICAL (Magnetic flux density)
Dimensions: M/CT

MULTIPLY	BY		TO OBTAIN
CGS e.m. units	*	1.0	gausses
CGS e.m. units		0.000 10	teslas
CGS e.s. units		2.9979×10^{10}	gausses
CGS e.s. units	**	$2.997\ 92 \times 10^{6}$	teslas
gausses		0.000 10	teslas
gausses		1.0×10^{-8}	webers/sq centimetre
gausses		0.000 10	webers/square metre
gausses (International)	**	1.000 331 0	gausses
gausses (International)		0.000 100 033	teslas
lines/square centimetre		1.0	gausses
lines/square centimetre		0.000 10	teslas
lines/square centimetre		1.0×10^{-8}	webers/sq centimetre
lines/square centimetre		6.4516×10^{-8}	webers/square inch
lines/square centimetre		0.000 10	webers/square metre
lines/square inch		0.155 000 310	gausses
lines/square inch		1.5500×10^{-5}	teslas
lines/square inch		1.5500×10^{-8}	webers/sq centimetre
lines/square inch		1.0×10^{-8}	webers/square inch
lines/square inch		1.5500×10^{-5}	webers/square metre
maxwells/sq centimetre	*	1.0	gausses
maxwells/sq centimetre		0.000 10	teslas
maxwells/sq centimetre		1.0×10^{-8}	webers/sq centimetre

Electrical (Magnetic flux density)

MULTIPLY	BY	TO OBTAIN
maxwells/sq centimetre	6.4516×10^{-8}	webers/square inch
maxwells/sq centimetre	0.000 10	webers/square metre
maxwells/square inch	0.155 000 310	gausses
maxwells/square inch	1.5500×10^{-5}	teslas
maxwells/square inch	1.5500×10^{-8}	webers/sq centimetre
maxwells/square inch	1.0×10^{-8}	webers/square inch
maxwells/square inch	1.5500×10^{-5}	webers/square metre
maxwells/square metre	1.0	teslas
teslas	* 10 000.0	gausses
teslas	0.000 10	webers/sq centimetre
teslas	0.000 645 160	webers/sq inch
teslas	* 1.0	webers/sq metre
webers/sq centimetre	1.0×10^{8}	gausses
webers/sq centimetre	10 000.0	teslas
webers/sq centimetre	6.451 60	webers/square inch
webers/sq centimetre	10 000.0	webers/square metre
webers/square inch	$1.550\ 00 \times 10^{7}$	gausses
webers/square inch	1550.003 100	teslas
webers/square inch	0.155 000 310	webers/sq centimetre
webers/square inch	1550.003 100	webers/square metre
webers/square metre	10 000.0	gausses
webers/square metre	1.0	teslas
webers/square metre	0.0001	webers/sq centimetre
webers/square metre	0.000 645 160	webers/square inch

ELECTRICAL (Magnetic flux)
Dimensions: ML^2/CT

MULTIPLY	BY	TO OBTAIN
CGS e.m. units	1.0×10^{-8}	tesla square metres
CGS e.m. units	* 1.0×10^{-8}	webers
CGS e.s. units	$2.997\ 92 \times 10^{2}$	tesla square metres
CGS e.s. units	** $2.997\ 92 \times 10^{2}$	webers
kilolines	1000.0	maxwells
kilolines	1.0×10^{-5}	webers
lines	* 1.0	maxwells
lines	1.0×10^{-8}	webers
maxwells	1.0	gauss square centimetres
maxwells	0.0010	kilolines
maxwells	0.000 10	tesla square centimetres

23

Electrical (Magnetic flux)

MULTIPLY	BY		TO OBTAIN
maxwells		1.0×10^{-8}	tesla square metres
maxwells	*	1.0×10^{-8}	volt seconds
maxwells	*	1.0×10^{-8}	webers
maxwells (International)	**	1.000 331 0	maxwells
megalines		1.0×10^{6}	maxwells
MKS (nr) units		12.533 670 54	webers
MKS (r) units		1.0	webers
teslas		1.0	webers/square metre
webers		1.0×10^{8}	maxwells
webers		1.0	tesla square metres
webers		1.0	volt seconds

ELECTRICAL (Magnetomotive force)
Dimensions: CL/T

MULTIPLY	BY		TO OBTAIN
Abampere turns		10.0	ampere turns
Abampere turns		10 000.0	milliampere turns
CGS e.m. units		0.795 774 715	ampere turns
CGS e.m. units		1.0	gilberts
CGS e.m. units	**	795.774 715 1	milliampere turns
CGS e.s. units		2.654×10^{-11}	ampere turns
CGS e.s. units	**	2.6544×10^{-8}	milliampere turns
gilberts		0.795 774 715	ampere turns
gilberts	**	795.774 715 1	millampere turns

ELECTRICAL (Permeability)
Dimensions: ML/C²

MULTIPLY	BY		TO OBTAIN
CGS e.m. units		1.2566×10^{-8}	henries/centimetre
CGS e.m. units	**	1.2566×10^{-6}	henries/metre
CGS e.s. units		1.1294×10^{13}	henries/centimetre
CGS e.s. units	**	1.1294×10^{15}	henries/metre
gausses/oersted		1.2566×10^{-8}	henries/centimetre
gausses/oersted		1.2566×10^{-6}	henries/metre
gausses/oersted	**	1.256 637 062	microhenries/metre
gausses/oersted		0.001 256 637	millihenries/metre
henries/centimetre		100.0	henries/metre
henries/centimetre		1.0×10^{8}	microhenries/metre

Electrical (Permeability)

MULTIPLY	BY	TO OBTAIN
henries/centimetre	100 000.0	millihenries/metre
henries/metre	10 000.0	microhenries/centimetre
henries/metre	10.0	millihenries/centimetre
microhenries/centimetre	0.000 10	henries/metre
microhenries/metre	1.0×10^{-8}	henries/centimetre
microhenries/metre	1.0×10^{-6}	henries/metre
millihenries/centimetre	0.10	henries/metre
millihenries/metre	1.0×10^{-5}	henries/centimetre
millihenries/metre	0.0010	henries/metre
MKS (nr) units	0.125 336 705 4	henries/centimetre
MKS (nr) units	12.533 670 54	henries/metre
MKS (r) units	0.010	henries/centimetre
MKS (r) units	1.0	henries/metre

ELECTRICAL (Resistance)
Dimensions: ML^2/C^2T

MULTIPLY		BY	TO OBTAIN
Abohms		1.0×10^{-15}	megohms
Abohms		0.0010	microhms
Abohms	*	1.0×10^{-9}	ohms
Abohms		1.113×10^{-21}	Statohms
CGS e.m. units		1.0	Abohms
CGS e.m. units	*	1.0×10^{-9}	ohms
CGS e.s. units	**	8.8975×10^{11}	ohms
CGS e.s. units		1.0	Statohms
megohms		1.0×10^{12}	microhms
megohms		1.0×10^{6}	ohms
microhms		1000.0	Abohms
microhms		1.0×10^{-12}	megohms
microhms		1.0×10^{-6}	ohms
microhms		1.113×10^{-18}	Statohms
ohms		1.0×10^{9}	Abohms
ohms		1.0×10^{-6}	megohms
ohms		1.0×10^{6}	microhms
ohms		0.999 505 245	ohms (International)
ohms		1.113×10^{-12}	Statohms
Statohms		8.9875×10^{20}	Abohms
Statohms		898 752.4324	megohms
Statohms		8.9875×10^{11}	ohms

ELECTRICAL (Resistivity)
Dimensions: ML^3/C^2T

MULTIPLY	BY	TO OBTAIN
Abohm centimetres	1.0×10^{-9}	ohm centimetres
Abohm centimetres	1.0×10^{-11}	ohm metres
CGS e.m. units	1.0	Abohm centimetres
CGS e.m. units	* 1.0×10^{-11}	ohm metres
CGS e.s. units	** 8987.524 324	megohm metres
CGS e.s. units	$8.987\,52 \times 10^9$	ohm metres
circular mil ohms/foot	** 166.242 611 3	Abohm centimetres
circular mil ohms/foot	0.166 242 611	microhm centimetres
circular mil ohms/foot	0.065 449 847	microhm inches
circular mil ohms/foot	1.6624×10^{-7}	ohm centimetres
microhm centimetres	1000.0	Abohm centimetres
microhm centimetres	0.393 700 787	microhm inches
microhm centimetres	1.0×10^{-6}	ohm centimetres
microhm inches	2540.0	Abohm centimetres
microhm inches	2.540	microhm centimetres
MKS units	1.0	ohm metres
ohm centimetres	1.0×10^6	microhm centimetres
ohm inches	2.540	ohm centimetres
ohm metres	1.0×10^{11}	Abohm centimetres
Statohm centimetres	$8.987\,52 \times 10^9$	ohm metres

ELECTRICAL (Surface charge density)
Dimensions: C/L^2

MULTIPLY	BY	TO OBTAIN
Abcoulombs/square cm	6.451 60	Abcoulombs/square inc
Abcoulombs/square cm	10 000.0	Abcoulombs/sq metre
Abcoulombs/square cm	10.0	coulombs/sq centimetre
Abcoulombs/square cm	64.5160	coulombs/square inch
Abcoulombs/square cm	100 000.0	coulombs/square metre
Abcoulombs/square inch	0.155 000 310	Abcoulombs/square cm
Abcoulombs/square inch	1550.003 100	Abcoulombs/sq metre
Abcoulombs/square inch	1.550 003 100	coulombs/sq centimetre
Abcoulombs/square inch	10.0	coulombs/square inch
Abcoulombs/square inch	15 500.031 00	coulombs/square metre
Abcoulombs/sq metre	0.000 10	Abcoulombs/square cm
Abcoulombs/sq metre	0.000 645 160	Abcoulombs/square inc

Electrical (Surface charge density)

MULTIPLY	BY	TO OBTAIN
Abcoulombs/sq metre	0.0010	coulombs/sq centimetre
Abcoulombs/sq metre	0.006 451 600	coulombs/square inch
Abcoulombs/sq metre	10.0	coulombs/square metre
CGS units of s.ch.d.	1.0	coulombs/sq centimetre
coulombs/sq centimetre	0.10	Abcoulombs/square cm
coulombs/sq centimetre	0.645 160	Abcoulombs/square inch
coulombs/sq centimetre	1000.0	Abcoulombs/sq metre
coulombs/sq centimetre	6.451 60	coulombs/square inch
coulombs/sq centimetre	10 000.0	coulombs/square metre
coulombs/square inch	0.015 500 031	Abcoulombs/square cm
coulombs/square inch	0.10	Abcoulombs/square inch
coulombs/square inch	155.000 310 0	Abcoulombs/sq metre
coulombs/square inch	0.155 000 310	coulombs/sq centimetre
coulombs/square inch	1550.003 100	coulombs/square metre
coulombs/square metre	0.000 010	Abcoulombs/square cm
coulombs/square metre	6.4516×10^{-5}	Abcoulombs/square inch
coulombs/square metre	0.10	Abcoulombs/sq metre
coulombs/square metre	0.000 10	coulombs/sq centimetre
coulombs/square metre	0.000 645 160	coulombs/square inch

ELECTRICAL (Surface current density)
Dimensions: C/TL^2

MULTIPLY	BY	TO OBTAIN
Abamperes/square cm	5.0671×10^{-5}	amperes/circular mil
Abamperes/square cm	10.0	amperes/sq centimetre
Abamperes/square cm	64.5160	amperes/square inch
amperes/sq centimetre	6.451 60	amperes/square inch
amperes/sq centimetre	10 000.0	amperes/square metre
amperes/square inch	1550.003 100	amperes/square metre
amperes/square inch	0.155 000 310	amps/sq centimetre
amperes/square metre	0.000 10	amperes/sq centimetre
amperes/square metre	0.000 645 160	amperes/square inch
amperes/square mil	15 500.031 00	Abamperes/square cm
amperes/square mil	155 000.3100	amperes/sq centimetre
amperes/square mil	1.550×10^9	amperes/square metre
CGS e.m. units	* 1.0×10^5	amperes/square metre
CGS e.s. units	3.3356×10^{-6}	amperes/square metre
MKS units	1.0	amperes/square metre

ELECTRICAL (Volume charge density)
Dimensions: C/L³

MULTIPLY	BY	TO OBTAIN
Abcoulombs/cubic cm	16.387 064 00	Abcoulombs/cubic inch
Abcoulombs/cubic cm	1.0×10^6	Abcoulombs/cubic metre
Abcoulombs/cubic cm	10.0	coulombs/cubic cm
Abcoulombs/cubic cm	163.870 640 0	coulombs/cubic inch
Abcoulombs/cubic cm	1.0×10^7	coulombs/cubic metre
Abcoulombs/cubic inch	0.061 023 744	Abcoulombs/cubic cm
Abcoulombs/cubic inch	61 023.744 09	Abcoulombs/cubic metre
Abcoulombs/cubic inch	0.610 237 441	coulombs/cubic cm
Abcoulombs/cubic inch	10.0	coulombs/cubic inch
Abcoulombs/cubic inch	610 237.4409	coulombs/cubic metre
Abcoulombs/cubic metre	1.0×10^{-6}	Abcoulombs/cubic cm
Abcoulombs/cubic metre	1.6387×10^{-5}	Abcoulombs/cubic inch
Abcoulombs/cubic metre	0.000 010	coulombs/cubic cm
Abcoulombs/cubic metre	0.000 163 871	coulombs/cubic inch
Abcoulombs/cubic metre	10.0	coulombs/cubic metre
CGS e.m. units	* 1.0×10^7	coulombs/cubic metre
CGS e.s. units	0.000 333 563	coulombs/cubic metre
coulombs/cubic cm	0.10	Abcoulombs/cubic cm
coulombs/cubic cm	1.638 706 400	Abcoulombs/cubic inch
coulombs/cubic cm	100 000.0	Abcoulombs/cubic metre
coulombs/cubic cm	16.387 064 00	coulombs/cubic inch
coulombs/cubic cm	1.0×10^6	coulombs/cubic metre
coulombs/cubic inch	0.006 102 374	Abcoulombs/cubic cm
coulombs/cubic inch	0.10	Abcoulombs/cubic inch
coulombs/cubic inch	6102.374 409	Abcoulombs/cubic metre
coulombs/cubic inch	0.061 023 744	coulombs/cubic cm
coulombs/cubic inch	61 023.744 09	coulombs/cubic metre
coulombs/cubic metre	1.0×10^{-7}	Abcoulombs/cubic cm
coulombs/cubic metre	1.6387×10^{-6}	Abcoulombs/cubic inch
coulombs/cubic metre	0.10	Abcoulombs/cubic metre
coulombs/cubic metre	1.0×10^{-6}	coulombs/cubic cm
coulombs/cubic metre	1.6387×10^{-5}	coulombs/cubic inch

ENERGY (Heat, work, electrical)
Dimensions: ML²/T²

MULTIPLY	BY	TO OBTAIN
atomic mass units (chem)	$9.313\ 95 \times 10^8$	electronvolts
atomic mass units (phys)	$9.311\ 41 \times 10^8$	electronvolts

Energy (Heat, work, electrical)

MULTIPLY	BY	TO OBTAIN
Btu	251.995 763 6	calories
Btu	0.555 555 556	Chu
Btu	1.0551×10^{10}	ergs
Btu	25 036.855 88	foot poundals
Btu	778.169 270 0	foot pounds(force)
Btu	$1.075\ 86 \times 10^{7}$	gram(force) centimetres
Btu	0.000 398 466	horsepower h (metric)
Btu	0.000 393 015	horsepower hours (UK)
Btu	4.4885×10^{-8}	horsepower years (UK)
Btu	4.5487×10^{-8}	horsepower yr (metric)
Btu	1055.055 863	joules (newton metres)
Btu	0.251 995 764	kilocalories
Btu	107.585 756 9	kilogram(force) metres
Btu	1.055 055 863	kilojoules
Btu	0.000 293 071	kilowatt hours
Btu	10.412 591 79	litre atmospheres
Btu	1.0×10^{-5}	therms
Btu	1055.055 863	watt seconds
Btu(15)	1054.728 269	joules (newton metres)
calories	0.003 968 321	Btu
calories	4.1868×10^{7}	ergs
calories	99.354 272 93	foot poundals
calories	3.088 025 206	foot pounds(force)
calories	42 693.478 40	gram(force) centimetres
calories	1.5812×10^{-6}	horsepower h (metric)
calories	1.5596×10^{-6}	horsepower hours (UK)
calories	* 4.186 80	joules (newton metres)
calories	0.0010	kilocalories
calories	0.426 934 784	kilogram(force) metres
calories	1.1630×10^{-6}	kilowatt hours
calories	0.041 320 503	litre atmospheres
calories	0.001 163 0	watt hours
calories	4.186 80	watt seconds
calories(15)	** 4.185 50	joules (newton metres)
cheval-vapeur heures	1.0	horsepower h (metric)
cheval-vapeur heures	$2.647\ 80 \times 10^{6}$	joules (newton metres)
Chu	1.80	Btu
Chu	453.592 374 5	calories
Chu	1.8991×10^{10}	ergs
Chu	45 066.340 59	foot poundals
Chu	1400.704 686	foot pounds(force)

29

Energy (Heat, work, electrical)

MULTIPLY	BY	TO OBTAIN
Chu	0.000 717 238	horsepower h (metric)
Chu	0.000 707 427	horsepower h (UK)
Chu	1899.100 553	joules (newton metres)
Chu	0.453 592 374	kilocalories
Chu	193.654 362 4	kilogram(force) metres
Chu	1.899 100 53	kilojoules
Chu	0.000 527 528	kilowatt hours
Chu	18.742 665 22	litre atmospheres
dyne centimetres	1.0	ergs
dyne centimetres	2.3730×10^{-6}	foot poundals
dyne centimetres	7.3756×10^{-8}	foot pounds(force)
dyne centimetres	0.001 019 716	gram(force) centimetres
dyne centimetres	8.8508×10^{-7}	inch pounds
dyne centimetres	* 1.0×10^{-7}	joules (newton metres)
dyne centimetres	1.0197×10^{-8}	kilogram(force) metres
electronvolts	1.602×10^{-12}	ergs
electronvolts	1.602×10^{-19}	joules (newton metres)
ergs	9.478×10^{-11}	Btu
ergs	2.3885×10^{-8}	calories
ergs	1.0	dyne centimetres
ergs	6.2422×10^{11}	electronvolts
ergs	2.3730×10^{-6}	foot poundals
ergs	7.3756×10^{-8}	foot pounds(force)
ergs	0.001 019 716	gram(force) centimetres
ergs	3.777×10^{-14}	horsepower h (metric)
ergs	3.725×10^{-14}	horsepower hours (UK)
ergs	* 1.0×10^{-7}	joules (newton metres)
ergs	2.389×10^{-11}	kilocalories
ergs	1.0197×10^{-8}	kilogram(force) metres
ergs	2.778×10^{-14}	kilowatt hours
ergs	9.869×10^{-10}	litre atmospheres
ergs	0.000 10	millijoules
ergs	2.778×10^{-11}	watt hours
ergs	1.0×10^{-7}	watt seconds
foot poundals	3.9940×10^{-5}	Btu
foot poundals	0.010 064 992	calories
foot poundals	421 401.1008	dyne centimetres
foot poundals	421 401.1008	ergs
foot poundals	0.031 080 950	foot pounds(force)
foot poundals	1.5915×10^{-8}	horsepower h (metric)
foot poundals	1.5697×10^{-8}	horsepower hours (UK)

Energy (Heat, work, electrical)

MULTIPLY	BY	TO OBTAIN
foot poundals	0.042 140 110	joules (newton metres)
foot poundals	0.004 297 095	kilogram(force) metres
foot poundals	1.1706×10^{-8}	kilowatt hours
foot poundals	0.000 415 891	litre atmospheres
foot pounds(force)	0.001 285 067	Btu
foot pounds(force)	0.323 831 554	calories
foot pounds(force)	0.000 713 926	Chu
foot pounds(force)	$1.355\ 82 \times 10^{7}$	dyne centimetres
foot pounds(force)	$1.355\ 82 \times 10^{7}$	ergs
foot pounds(force)	32.174 048 56	foot poundals
foot pounds(force)	13 825.495 43	gram(force) centimetres
foot pounds(force)	5.1206×10^{-7}	horsepower h (metric)
foot pounds(force)	5.0505×10^{-7}	horsepower hours (UK)
foot pounds(force)	1.355 817 948	joules (newton metres)
foot pounds(force)	0.000 323 832	kilocalories
foot pounds(force)	0.138 254 954	kilogram(force) metres
foot pounds(force)	3.7662×10^{-7}	kilowatt hours
foot pounds(force)	0.013 380 883	litre atmospheres
foot pounds(force)	0.000 376 616	watt hours
gram(force) centimetres	9.2949×10^{-8}	Btu
gram(force) centimetres	2.3423×10^{-5}	calories
gram(force) centimetres	980.6650	dyne centimetres
gram(force) centimetres	980.6650	ergs
gram(force) centimetres	0.002 327 153	foot poundals
gram(force) centimetres	7.2330×10^{-5}	foot pounds(force)
gram(force) centimetres	3.704×10^{-11}	horsepower hrs (metric)
gram(force) centimetres	3.653×10^{-11}	horsepower hours (UK)
gram(force) centimetres	9.8067×10^{-5}	joules (newton metres)
gram(force) centimetres	2.3423×10^{-8}	kilocalories
gram(force) centimetres	1.0×10^{-5}	kilogram(force) metres
gram(force) centimetres	2.724×10^{-11}	kilowatt hours
gram(force) centimetres	2.7241×10^{-8}	watt hours
horsepower h (metric)	2509.625 881	Btu
horsepower h (metric)	$6.324\ 15 \times 10^{5}$	calories
horsepower h (metric)	1.0	cheval-vapeur heures
horsepower h (metric)	2.6478×10^{13}	ergs
horsepower h (metric)	$1.952\ 91 \times 10^{6}$	foot pounds(force)
horsepower h (metric)	0.986 320 071	horsepower hours (UK)
horsepower h (metric)	$2.647\ 80 \times 10^{6}$	joules (newton metres)
horsepower h (metric)	632.415 090 3	kilocalories
horsepower h (metric)	$2.700\ 00 \times 10^{5}$	kilogram(force) metres

31

Energy (Heat, work, electrical)

MULTIPLY	BY	TO OBTAIN
horsepower h (metric)	0.735 498 750	kilowatt hours
horsepower h (metric)	735.498 750 0	watt hours
horsepower hours (UK)	2544.433 552	Btu
horsepower hours (UK)	641 186.4758	calories
horsepower hours (UK)	2.6845×10^{13}	ergs
horsepower hours (UK)	$1.980\ 00 \times 10^{6}$	foot pounds(force)
horsepower hours (UK)	1.013 869 665	horsepower h (metric)
horsepower hours (UK)	$2.684\ 52 \times 10^{6}$	joules (newton metres)
horsepower hours (UK)	641.186 475 8	kilocalories
horsepower hours (UK)	273 744.8096	kilogram(force) metres
horsepower hours (UK)	0.745 699 871	kilowatt hours
horsepower hours (UK)	745.699 871 4	watt hours
joules (newton metres)	0.000 947 817	Btu
joules (newton metres)	0.238 845 897	calories
joules (newton metres)	1.0×10^{7}	ergs
joules (newton metres)	23.730 360 41	foot poundals
joules (newton metres)	0.737 562 150	foot pounds(force)
joules (newton metres)	10 197.162 13	gram(force) centimetres
joules (newton metres)	3.7251×10^{-7}	horsepower hours (UK)
joules (newton metres)	3.7767×10^{-7}	horsepower h (metric)
joules (newton metres)	0.000 238 846	kilocalories
joules (newton metres)	0.101 971 621	kilogram(force) metres
joules (newton metres)	2.7778×10^{-7}	kilowatt hours
joules (newton metres)	0.009 869 233	litre atmospheres
joules (newton metres)	9.4782×10^{-9}	therms
joules (newton metres)	1/3600	watt hours
joules (newton metres)	1.0	watt seconds
kilocalories	3.968 320 680	Btu
kilocalories	1000.0	calories
kilocalories	4.1868×10^{10}	ergs
kilocalories	99 354.272 96	foot poundals
kilocalories	3088.025 207	foot pounds(force)
kilocalories	0.001 581 240	horsepower h (metric)
kilocalories	0.001 559 609	horsepower hours (UK)
kilocalories	4186.80	joules (newton metres)
kilocalories	1.0	kilocalories
kilocalories	426.934 784 0	kilogram(force) metres
kilocalories	0.001 163 0	kilowatt hours
kilocalories	41.320 503 33	litre atmospheres
kilocalories	1.1630	watt hours
kilogram(force) metres	0.009 294 911	Btu

Energy (Heat, work, electrical)

MULTIPLY	BY	TO OBTAIN
kilogram(force) metres	2.342 278 112	calories
kilogram(force) metres	$9.806\ 65 \times 10^7$	dyne centimetres
kilogram(force) metres	$9.806\ 65 \times 10^7$	ergs
kilogram(force) metres	232.715 338 9	foot poundals
kilogram(force) metres	7.233 013 853	foot pounds(force)
kilogram(force) metres	1.0×10^5	gram(force) centimetres
kilogram(force) metres	3.7037×10^{-6}	horsepower h (metric)
kilogram(force) metres	3.6530×10^{-6}	horsepower hours (UK)
kilogram(force) metres	9.806 650	joules (newton metres)
kilogram(force) metres	0.002 342 278	kilocalories
kilogram(force) metres	2.7241×10^{-6}	kilowatt hours
kilogram(force) metres	0.096 784 111	litre atmospheres
kilogram(force) metres	7.233 013 853	pounds feet
kilogram(force) metres	0.002 724 069	watt hours
kilograms of ice melted	** 334.0	kilojoules (latent heat)
kilojoules	1000.0	joules (newton metres)
kilowatt hours	3412.141 600	Btu
kilowatt hours	859 845.2279	calories
kilowatt hours	3.6000×10^{13}	ergs
kilowatt hours	8.5429×10^7	foot poundals
kilowatt hours	$2.655\ 22 \times 10^6$	foot pounds(force)
kilowatt hours	1.359 621 617	horsepower hrs (metric)
kilowatt hours	1.341 022 090	horsepower hours (UK)
kilowatt hours	3.60×10^6	joules (newton metres)
kilowatt hours	859.845 227 9	kilocalories
kilowatt hours	367 097.8367	kilogram(force) metres
kilowatt hours	3600.0	kilojoules
kilowatt hours	22.728 068 04	lb water (62–212 °F)
kilowatt hours	3.516 923 542	lb water (evap 100 °C)
kilowatt hours	3.600	megajoules
kilowatt hours	1000.0	watt hours
litre atmospheres	0.096 037 569	Btu
litre atmospheres	24.201 060 48	calories
litre atmospheres	0.035 314 667	cubic feet atmospheres
litre atmospheres	2404.478 769	foot poundals
litre atmospheres	74.733 484 79	foot pounds(force)
litre atmospheres	3.8268×10^{-5}	horsepower h (metric)
litre atmospheres	3.7744×10^{-5}	horsepower hours (UK)
litre atmospheres	* 101.3250	joules (newton metres)
litre atmospheres	10.332 274 53	kilogram(force) metres
litre atmospheres	2.8146×10^{-5}	kilowatt hours

33

Energy (Heat, work, electrical)

MULTIPLY	BY	TO OBTAIN
myriawatts	10.0	kilowatts
thermies	** $4.185\ 50 \times 10^6$	joules (newton metres)
thermies	$4.185\ 50$	megajoules
therms	* 1.0×10^5	Btu
therms	$2.519\ 96 \times 10^7$	calories
therms	1.0551×10^{15}	ergs
therms	$2.503\ 69 \times 10^9$	foot poundals
therms	$7.781\ 69 \times 10^7$	foot pounds(force)
therms	$39.846\ 576\ 64$	horsepower h (metric)
therms	$39.301\ 478\ 28$	horsepower hours (UK)
therms	$1.055\ 06 \times 10^8$	joules (newton metres)
therms	$1.055\ 06 \times 10^5$	kilojoules
therms	$105.505\ 586\ 3$	megajoules
watt hours	$3.412\ 141\ 60$	Btu
watt hours	$859.845\ 227\ 9$	calories
watt hours	3.6000×10^{10}	ergs
watt hours	$2655.223\ 738$	foot pounds(force)
watt hours	$0.001\ 341\ 022$	horsepower hours (UK)
watt hours	$0.001\ 359\ 622$	horsepower h (metric)
watt hours	3600.0	joules (newton metres)
watt hours	$0.859\ 845\ 228$	kilocalories
watt hours	$367.097\ 836\ 7$	kilogram(force) metres
watt hours	0.0010	kilowatt hours
watt seconds	$0.737\ 562\ 150$	foot pounds(force)
watt seconds	$10\ 197.162\ 13$	gram(force) centimetres
watt seconds	1.0	joules (newton metres)
watt seconds	$0.009\ 869\ 233$	litre atmospheres
watt seconds	1.0	volt coulombs

ENTHALPY
Dimensions: L^2/T^2

MULTIPLY	BY	TO OBTAIN
Btu/pound	$0.555\ 555\ 556$	calories/gram
Btu/pound	$778.169\ 270\ 1$	foot pounds(force)/lb
Btu/pound	$0.000\ 393\ 015$	horsepower h (UK)/lb
Btu/pound	2.3260	joules/gram
Btu/pound	2326.0	joules/kilogram
Btu/pound	$0.555\ 555\ 556$	kg calories/kilogram
Btu/pound	2.3260	kilojoules/kilogram
calories/gram	1.80	Btu/pound

Enthalpy

MULTIPLY	BY	TO OBTAIN
calories/gram	1400.704 686	foot pounds(force)/lb
calories/gram	4.186 80	joules/gram
calories/gram	4186.80	joules/kilogram
calories/gram	4.186 80	kilojoules/kilogram
calories/gram	0.001 163 0	watt hours/gram
Chu/pound	1.80	Btu/pound
Chu/pound	1.0	calories/gram
Chu/pound	1400.704 686	foot pounds(force)/lb
Chu/pound	4.186 80	joules/gram
Chu/pound	4186.8	joules/kilogram
Chu/pound	4.1868	kilojoules/kilogram
foot pounds(force)/lb	0.001 285 067	Btu/pound
foot pounds(force)/lb	0.000 713 926	calories/gram
foot pounds(force)/lb	0.002 989 067	joules/gram
foot pounds(force)/lb	0.000 304 800	kg(force) metres/gram
foot pounds(force)/lb	0.002 989 067	kilojoules/kilogram
foot pounds(force)/lb	8.303×10^{-10}	kilowatt hours/gram
joules/gram	0.429 942 261 4	Btu/pound
joules/gram	1.0	kilojoules/kilogram
joules/kilogram	0.0010	kilojoules/kilogram
kilocalories/kilogram	1.80	Btu/pound
kilocalories/kilogram	4186.80	joules/kilogram
kilocalories/kilogram	4.186 80	kilojoules/kilogram
kilojoules/kilogram	1000.0	joules/kilogram
kilowatt hours/gram	$1.547\ 72 \times 10^{6}$	Btu/pound
kilowatt hours/gram	859 845.2279	calories/gram
kilowatt hours/gram	569 123.7970	cubic feet atm/pound
kilowatt hours/gram	608.277 394 1	horsepower h (UK)/lb
kilowatt hours/gram	3.60×10^{6}	joules/gram

ENTROPY
Dimensions: L^2/T^2

MULTIPLY	BY	TO OBTAIN
Btu/pound °F	4186.80	joules/kilogram K
Btu/pound °F	4.186 80	kilojoules/kg K
Btu/pound °R	4186.80	joules/kilogram K
Btu/pound °R	4.186 80	kilojoules/kg K
calories/gram °C	4186.80	joules/kilogram K
calories/gram °C	4.186 80	kilojoules/kg K
Chu/pound °C	4186.80	joules/kilogram K

Entropy

MULTIPLY	BY	TO OBTAIN
Chu/pound °C	4.186 80	kilojoules/kg K
joules/kilogram °C	1.0	joules/kilogram K
joules/kilogram °C	0.0010	kilojoules/kg K
kilocalories/kg °C	4186.80	joules/kilogram K
kilocalories/kg °C	4.186 80	kilojoules/kg K
kilojoules/kg °C	1000.0	joules/kilogram K

FORCE (Weight)
Dimensions: ML/T²

MULTIPLY	BY		TO OBTAIN
carats (metric)	*	200.0	milligrams
centals		45.359 237 45	kilograms
centals	*	100.0	pounds (avoirdupois)
drams (apothecary/troy)		2.194 285 714	drams (avoirdupois)
drams (apothecary/troy)	*	60.0	grains
drams (apothecary/troy)		3.887 934 637	grams
drams (apothecary/troy)		0.137 142 857	ounces (avoirdupois)
drams (apothecary/troy)		0.1250	ounces (troy)
drams (apothecary/troy)	*	3.0	scruples (apothecary)
drams (avoirdupois)		0.445 729 167	drams (apothecary/troy)
drams (avoirdupois)		27.343 750	grains
drams (avoirdupois)		1.771 845 213	grams
drams (avoirdupois)		0.062 50	ounces (avoirdupois)
drams (avoirdupois)		0.056 966 146	ounces (apothecary/troy)
drams (avoirdupois)		1.139 322 917	pennyweights
drams (avoirdupois)		0.003 906 250	pounds (avoirdupois)
drams (avoirdupois)		0.004 747 179	pounds (apothecary/troy)
drams (avoirdupois)		1.367 187 50	scruples (apothecary)
dynes		0.015 736 626	grains
dynes		0.001 019 716	grams
dynes		1.0×10^{-7}	joules/centimetre
dynes		1.0197×10^{-6}	kilograms
dynes	*	1.0×10^{-5}	newtons (joules/metre)
dynes		7.2330×10^{-5}	poundals
dynes		2.2481×10^{-6}	pounds (avoirdupois)
grains		0.323 994 550	carats (metric)
grains		1/60	drams (apothecary/troy)
grains		0.036 571 429	drams (avoirdupois)
grains		63.546 023 71	dynes
grains	*	1.0	grains (apothecary)

Force (Weight)

MULTIPLY	BY	TO OBTAIN
grains	* 1.0	grains (avoirdupois)
grains	* 1.0	grains (troy)
grains	0.064 798 911	grams
grains	6.4799×10^{-5}	kilograms
grains	64.798 910 64	milligrams
grains	0.002 285 714	ounces (avoirdupois)
grains	1/480	ounces (apothecary/troy)
grains	1/24	pennyweight (troy)
grains	1/7000	pounds (avoirdupois)
grains	1/5760	pounds (apothecary/troy)
grains	0.050	scruples (apothecary)
grains	6.4799×10^{-8}	tonnes (metric)
hundredweights (long)	50.802 345 94	kilograms
hundredweights (long)	* 112.0	pounds (avoirdupois)
hundredweights (long)	0.050	tons (long)
hundredweights (long)	0.050 802 346	tons (metric)
hundredweights (long)	0.0560	tons (short)
hundredweights (short)	45.359 237 45	kilograms
hundredweights (short)	1600.0	ounces (avoirdupois)
hundredweights (short)	* 100.0	pounds (avoirdupois)
hundredweights (short)	0.044 642 857	tons (long)
hundredweights (short)	0.045 359 237	tons (metric)
hundredweights (short)	0.050	tons (short)
joules/centimetre	1.0×10^{7}	dynes
joules/centimetre	10 197.162 13	grams
joules/centimetre	100.0	newtons (joules/metre)
joules/centimetre	723.301 378 1	poundals
joules/centimetre	22.480 894 09	pounds (avoirdupois)
kilograms	15 432.358 20	grains
kilograms	1000.0	grams
kilograms	0.019 684 131	hundredweights (long)
kilograms	0.022 046 226	hundredweights (short)
kilograms	* 9.806 650	newtons (joules/metre)
kilograms	35.273 961 60	ounces (avoirdupois)
kilograms	2.204 622 60	pounds (avoirdupois)
kilograms	0.0010	tonnes (metric)
kilograms	0.000 984 207	tons (long)
kilonewtons	1000.0	newtons (joules/metre)
kiloponds (kps, Germany)	1.0	kilograms (force)
kilopounds (kips, US)	1000.0	pounds(force)
millinewtons	0.0010	newtons (joules/metre)

Force (Weight)

MULTIPLY	BY	TO OBTAIN
newtons (joules/metre)	1.0×10^5	dynes
newtons (joules/metre)	101.971 621 3	grams
newtons (joules/metre)	0.101 971 621	kilograms
newtons (joules/metre)	3.278 463 721	ounces (apothecary/troy)
newtons (joules/metre)	3.596 943 054	ounces (avoirdupois)
newtons (joules/metre)	7.233 013 781	poundals
newtons (joules/metre)	0.273 205 310	pounds (apothecary/troy)
newtons (joules/metre)	0.224 808 941	pounds (avoirdupois)
ounces (apothecary/troy)	3.110 347 711	decagrams
ounces (apothecary/troy)	* 8.0	drams (apothecary/troy)
ounces (apothecary/troy)	17.554 285 71	drams (avoirdupois)
ounces (apothecary/troy)	* 480.0	grains
ounces (apothecary/troy)	31.103 477 11	grams
ounces (apothecary/troy)	31 103.477 11	milligrams
ounces (apothecary/troy)	1.097 142 857	ounces (avoirdupois)
ounces (apothecary/troy)	* 20.0	pennyweights (troy)
ounces (apothecary/troy)	1/12	pounds (apothecary/troy)
ounces (apothecary/troy)	0.068 571 429	pounds (avoirdupois)
ounces (apothecary/troy)	* 24.0	scruples (apothecary)
ounces (apothecary/troy)	3.4286×10^{-5}	tons (short)
ounces (avoirdupois)	7.291 666 667	drams (apothecary/troy)
ounces (avoirdupois)	* 16.0	drams (avoirdupois)
ounces (avoirdupois)	437.50	grains
ounces (avoirdupois)	28.349 523 41	grams
ounces (avoirdupois)	0.000 558 036	hundredweights (long)
ounces (avoirdupois)	0.000 625 0	hundredweights (short)
ounces (avoirdupois)	0.278 013 854	joules/metre (newtons)
ounces (avoirdupois)	0.028 349 523	kilograms
ounces (avoirdupois)	0.278 013 854	newtons (joules/metre)
ounces (avoirdupois)	0.911 458 333	ounces (apothecary/troy)
ounces (avoirdupois)	18.229 166 67	pennyweights (troy)
ounces (avoirdupois)	0.075 954 861	pounds (apothecary/troy)
ounces (avoirdupois)	1/16	pounds (avoirdupois)
ounces (avoirdupois)	21.8750	scruples (apothecary)
ounces (avoirdupois)	2.8350×10^{-5}	tonnes (metric)
ounces (avoirdupois)	2.7902×10^{-5}	tons (long)
ounces (avoirdupois)	3.1250×10^{-5}	tons (short)
pennyweights (troy)	0.40	drams (apothecary/troy)
pennyweights (troy)	0.877 714 286	drams (avoirdupois)
pennyweights (troy)	* 24.0	grains
pennyweights (troy)	1.555 173 856	grams

Force (Weight)

MULTIPLY	BY	TO OBTAIN
pennyweights (troy)	0.054 857 143	ounces (avoirdupois)
pennyweights (troy)	0.050	ounces (apothecary/troy)
pennyweights (troy)	0.003 428 571	pounds (avoirdupois)
pennyweights (troy)	1/240	pounds (apothecary/troy)
poundals	13 825.495 57	dynes
poundals	14.098 081 99	grams
poundals	0.001 382 550	joules/centimetre
poundals	0.138 254 956	joules/metre (newtons)
poundals	0.014 098 082	kilograms
poundals	0.031 080 950	pounds (avoirdupois)
pounds (apothecary/troy)	96.0	drams (apothecary/troy)
pounds (apothecary/troy)	210.651 428 6	drams (avoirdupois)
pounds (apothecary/troy)	* 5760.0	grains
pounds (apothecary/troy)	373.241 725 3	grams
pounds (apothecary/troy)	0.373 241 725	kilograms
pounds (apothecary/troy)	* 12.0	ounces (apothecary/troy)
pounds (apothecary/troy)	13.165 714 29	ounces (avoirdupois)
pounds (apothecary/troy)	240.0	pennyweights (troy)
pounds (apothecary/troy)	0.822 857 143	pounds (avoirdupois)
pounds (apothecary/troy)	288.0	scruples (apothecary)
pounds (apothecary/troy)	0.000 367 347	tons (long)
pounds (apothecary/troy)	0.000 373 242	tons (metric)
pounds (apothecary/troy)	0.000 411 429	tons (short)
pounds (avoirdupois)	116.666 666 7	drams (apothecary/troy)
pounds (avoirdupois)	256.0	drams (avoirdupois)
pounds (avoirdupois)	444 822.1659	dynes
pounds (avoirdupois)	* 7000.0	grains
pounds (avoirdupois)	1/112	hundredweights (long)
pounds (avoirdupois)	0.010	hundredweights (short)
pounds (avoirdupois)	0.044 482 217	joules/centimetre
pounds (avoirdupois)	4.448 221 659	joules/metre (newtons)
pounds (avoirdupois)	0.453 592 375	kilograms
pounds (avoirdupois)	4.448 221 659	newtons (joules/metre)
pounds (avoirdupois)	14.583 333 334	ounces (apothecary/troy)
pounds (avoirdupois)	* 16.0	ounces (avoirdupois)
pounds (avoirdupois)	291.666 666 7	pennyweights (troy)
pounds (avoirdupois)	** 32.174 048 56	poundals
pounds (avoirdupois)	1.215 277 778	pounds (apothecary/troy)
pounds (avoirdupois)	350.0	scruples (apothecary)
pounds (avoirdupois)	0.000 453 592	tonnes (metric)
pounds (avoirdupois)	1/2240	tons (long)

Force (Weight)

MULTIPLY	BY	TO OBTAIN
pounds (avoirdupois)	0.000 50	tons (short)
quintals (metric)	100 000.0	grams
quintals (metric)	1.968 413 036	hundredweights (long)
quintals (metric)	* 100.0	kilograms
quintals (metric)	220.462 260	pounds (avoirdupois)
quintals (Imp, long)	1.0	hundredweights (long)
quintals (Imp, long)	1.120	hundredweights (short)
quintals (Imp, long)	50.802 345 94	kilograms
quintals (Imp, long)	* 112.0	pounds (avoirdupois)
quintals (Imp, short)	0.892 857 143	hundredweights (long)
quintals (Imp, short)	1.0	hundredweights (short)
quintals (Imp, short)	45.359 237 45	kilograms
quintals (Imp, short)	* 100.0	pounds (avoirdupois)
scruples (apothecary)	1/3	drams (apothecary/troy)
scruples (apothecary)	0.731 428 571	drams (avoirdupois)
scruples (apothecary)	* 20.0	grains
scruples (apothecary)	1.295 978 213	grams
scruples (apothecary)	1/24	ounces (apothecary/troy)
scruples (apothecary)	0.045 714 286	ounces (avoirdupois)
scruples (apothecary)	0.833 333 333	pennyweights (troy)
scruples (apothecary)	0.002 857 143	pounds (avoirdupois)
scruples (apothecary)	0.003 472 222	pounds (apothecary/troy)
stones	6.350 293 243	kilograms
stones	* 14.0	pounds (avoirdupois)
tonnes (metric)	$9.806\ 65 \times 10^8$	dynes
tonnes (metric)	19.684 130 36	hundredweights (long)
tonnes (metric)	22.046 226 00	hundredweights (short)
tonnes (metric)	* 1000.0	kilograms
tonnes (metric)	9.806 650	kilonewtons
tonnes (metric)	9806.650	newtons (joules/metre)
tonnes (metric)	35 273.961 60	ounces (avoirdupois)
tonnes (metric)	2204.622 600	pounds (avoirdupois)
tonnes (metric)	2679.228 854	pounds (apothecary/troy)
tonnes (metric)	0.984 206 518	tons (long)
tons (long)	9.9640×10^8	dynes
tons (long)	20.0	hundredweights (long)
tons (long)	22.40	hundredweights (short)
tons (long)	9.964 016 518	kilonewtons
tons (long)	9964.016 518	newtons (joules/metre)
tons (long)	35 840.0	ounces (avoirdupois)
tons (long)	* 2240.0	pounds (avoirdupois)

Force (Weight)

MULTIPLY	BY	TO OBTAIN
tons (long)	2722.222 222	pounds (apothecary/troy)
tons (long)	1.016 046 919	tonnes (metric)
tons (short)	$8.896\ 44 \times 10^8$	dynes
tons (short)	17.857 142 86	hundredweights (long)
tons (short)	20.0	hundredweights (short)
tons (short)	* 2000.0	pounds (avoirdupois)
tons (short)	0.907 184 749	tonnes (metric)
tons (short)	0.892 857 143	tons (long)

FORCE PER UNIT LENGTH (Spring rate)
Dimensions: M/T^2

MULTIPLY	BY	TO OBTAIN
grams/centimetre	980.6650	dynes/centimetre
grams/centimetre	2.540	grams/inch
grams/centimetre	100.0	kilograms/kilometre
grams/centimetre	0.10	kilograms/metre
grams/centimetre	0.180 166 352	poundals/inch
grams/centimetre	0.067 196 897	pounds/foot
grams/centimetre	0.005 599 741	pounds/inch
grams/centimetre	0.10	tonnes/kilometre
kilograms/metre	0.671 968 969	pounds/foot
newtons/millimetre	5.710 147 099	pounds/inch
pounds/foot	1.488 163 958	kilograms/metre
pounds/inch	178.579 675 0	grams/centimetre
pounds/inch	5443.108 494	grams/foot
pounds/inch	453.592 374 5	grams/inch
pounds/inch	0.175 126 837	newtons/millimetre
pounds/inch	6.299 212 598	ounces/centimetre
pounds/inch	16.0	ounces/inch
pounds/inch	39.370 078 74	pounds/metre
milligrams/inch	0.386 088 583	dynes/centimetre
milligrams/inch	0.980 665 0	dynes/inch
milligrams/inch	0.000 393 701	grams/centimetre
milligrams/inch	0.000 10	grams/inch
milligrams/millimetre	9.806 650	dynes/centimetre

HEAT CAPACITY
Dimensions: L^2/T^2

MULTIPLY	BY	TO OBTAIN
Btu/pound °C	* 2.3260	joules/gram °C
Btu/pound °F	4186.80	joules/kilogram K
Btu/pound °F	4.186 80	kilojoules/kg K
Btu/pound °R	4186.80	joules/kilogram K
Btu/pound °R	4.186 80	kilojoules/kg K
calories/gram °C	4186.80	joules/kilogram K
calories/gram °C	4.186 80	kilojoules/kg K
Chu/pound °C	4186.80	joules/kilogram K
Chu/pound °C	4.186 80	kilojoules/kg K
joules/kilogram °C	1.0000	joules/kilogram K
joules/kilogram °C	0.0010	kilojoules/kg K
kilocals/kilogram °C	4186.80	joules/kilogram K
kilocals/kilogram °C	4.186 80	kilojoules/kg K
kilojoules/kg °C	1000.0	joules/kilogram K

HEAT FLUX DENSITY
Dimensions: M/T^3

MULTIPLY	BY	TO OBTAIN
Btu/hour foot²	3.154 590 778	joules/second metre²
Btu/hour foot²	0.003 154 591	kilowatts/metre²
Btu/hour foot²	3.154 590 778	watts/metre²
Btu/min foot²	0.122 112 947	watts/square inch
cal/second centimetre²	41 868.0	joules/second metre²
cal/second centimetre²	41.8680	kilowatts/metre²
Chu/hour foot²	5.678 263 398	joules/second metre²
Chu/hour foot²	0.005 678 263	kilowatts/metre²
dynes/hour centimetre	0.010	ergs/hour millimetre²
kilocals/hour foot²	12.518 427 82	joules/second metre²
kilocals/hour foot²	0.012 518 428	kilowatts/metre²
kilocals/hour metre²	1.1630	joules/second metre²
kilocals/hour metre²	0.001 163 0	kilowatts/metre²
kilocals/hour metre²	1.1630	watts/metre²
kilowatts/metre²	1000.0	joules/second metre²
watts/centimetre²	3169.983 276	Btu/hour foot²
watts/centimetre²	41 113.059 52	foot pounds/min ft²
watts/centimetre²	859.845 227 8	gram cal/hour cm²
watts/metre²	1.0	joules/second metre²

Heat Flux Density

MULTIPLY	BY	TO OBTAIN
watts/metre^2	0.0010	kilowatts/metre^2
watts/square inch	491.348 390 4	Btu/hour foot^2
watts/square inch	6372.536 972	foot pounds/min ft^2
watts/square inch	133.276 276 9	gram cal/hour cm^2
watts/square inch	0.195 785 513	hp(metric)/foot^2
watts/square inch	0.193 107 181	hp(UK)/foot^2

HEAT TRANSFER COEFFICIENT
Dimensions: M/T^3

MULTIPLY	BY	TO OBTAIN
Btu/hour foot^2 °F	5.678 263 40	joules/s metre^2 K
Btu/hour foot^2 °F	5.678 263 40	watts/metre^2 K
cal/s cms^2 °C	41 868.0	joules/s metre^2 K
cal/s cms^2 °C	41 868.0	watts/metre^2 K
Chu/hour foot^2 °C	5.678 263 398	joules/s metre^2 K
Chu/hour foot^2 °C	5.678 263 398	watts/metre^2 K
joules/second metre^2 K	1.0	watts/metre^2 K
kilocal/hour foot^2 °C	12.518 427 82	joules/s metre^2 K
kilocal/hour foot^2 °C	12.518 427 82	watts/metre^2 K
kilocal/hour metre^2 °C	1.1630	joules/s metre^2 K
kilocal/hour metre^2 °C	1.1630	watts/metre^2 K
watts/metre^2 K	1.0	joules/s metre^2 K
watts/metre^2 °C	1.0	joules/s metre^2 K

ILLUMINATION (Luminous flux)
Dimensions: ML^2/T^3

MULTIPLY	BY	TO OBTAIN
candle power (spherical)	** 12.566 370 62	lumens
lumens	0.079 577 472	candle power (spherical)
lumens	** 0.001 470 588	watts

ILLUMINATION (Luminous incidence)
Dimensions: L^2/MT^3

MULTIPLY	BY	TO OBTAIN
candles/sq centimetre	6.451 60	candles/square inch
candles/sq centimetre	10 000.0	candles/square metre
candles/sq centimetre	2918.635 081	foot lamberts
candles/sq centimetre	3.141 592 654	lamberts
candles/sq centimetre	1/60	new candles
candles/square foot	1/144	candles/square inch

Illumination (Luminous incidence)

MULTIPLY	BY	TO OBTAIN
candles/square foot	10.763 91	candles/square metre
candles/square foot	3.141 592 654	foot lamberts
candles/square foot	0.003 381 582	lamberts
candles/square inch	0.155 000 310	candles/sq centimetre
candles/square inch	144.0	candles/square foot
candles/square inch	452.389 342 3	foot lamberts
candles/square inch	0.486 947 835	lamberts
foot candles	* 1.0	lumens/square foot
foot candles	10.763 910 42	lumens/square metre
foot candles	10.763 910 42	lux
foot candles	1.076 391 042	milliphots
foot lamberts	0.000 342 626	candles/sq centimetre
foot lamberts	0.318 309 886	candles/square foot
foot lamberts	0.001 076 391	lamberts
foot lamberts	1.0	lumens/square foot
foot lamberts	1.076 391 042	millilamberts
lamberts	0.318 309 886	candles/sq centimetre
lamberts	295.719 560 8	candles/square foot
lamberts	2.053 608 061	candles/square inch
lamberts	929.030 40	foot lamberts
lamberts	1.0	lumens/sq centimetre
lumens/sq centimetre	1.0	lamberts
lumens/sq centimetre	1.0	phots
lumens/square foot	1.0	foot candles
lumens/square foot	1.0	foot lamberts
lumens/square foot	10.763 910 42	lumens/square metre
lumens/square metre	0.092 903 040	foot candles
lumens/square metre	0.092 903 040	lumens/square foot
lumens/square metre	0.000 10	phots
lux	0.092 903 040	foot candles
lux	* 1.0	lumens/square metre
lux	0.000 10	phots
millilamberts	0.000 318 310	candles/sq centimetre
millilamberts	0.002 053 608	candles/square inch
millilamberts	0.929 030 40	foot lamberts
millilamberts	0.0010	lamberts
millilamberts	0.0010	lumens/sq centimetre
millilamberts	0.929 030 40	lumens/square foot
milliphots	0.929 030 40	foot candles
milliphots	0.929 030 40	lumens/square foot
milliphots	10.0	lumens/square metre

Illumination (Luminous incidence)

MULTIPLY	BY	TO OBTAIN
milliphots	10.0	lux
milliphots	0.0010	phots
phots	929.030 40	foot candles
phots	1.0	lumens/sq centimetre
phots	10 000.0	lumens/square metre
phots	10 000.0	lux
stilbs	1.0	candles/sq centimetre
stilbs	6.451 60	candles/square inch
stilbs	** 3.141 592 654	lamberts

ILLUMINATION (Luminous intensity)
Dimensions: ML^2/T^3

MULTIPLY	BY	TO OBTAIN
bougie decimales	1.0	candles (International)
bougie decimales	1.0	decimal candle
candles (German)	1.052 631 579	candles (International)
candles (German)	1.010 526 316	candles (UK)
candles (German)	1.169 590 643	hefner units
candles (International)	* 0.950	candles (German)
candles (International)	1.0	candles (pentane)
candles (International)	* 0.960	candles (UK)
candles (International)	0.104 058 273	carcel units
candles (International)	1.111 111 111	hefner units
candles (International)	1.0	lumens(Int)/steradian
candles (pentane)	1.0	candles (International)
candles (UK)	1.041 666 667	candles (International)
candles (UK)	1.157 407 408	hefner units
carcel units	* 9.610	candles (International)
decimal candle	1.0	bougie decimales
hefner candles	0.8550	candles (German)
hefner candles	* 0.90	candles (International)
hefner candles	0.8640	candles (UK)
hefner candles	0.090	pentane candles (10 cp)
new candles	* 60.0	candles/sq centimetre

LINEAR MEASURE (Distance or depth)
Dimensions: L

MULTIPLY	BY	TO OBTAIN
angstrom units	1.0×10^{-8}	centimetres
angstrom units	3.937×10^{-9}	inches

Linear Measure (Distance or depth)

MULTIPLY	BY	TO OBTAIN
angstrom units	* 1.0×10^{-10}	metres
angstrom units	0.000 10	microns (μ)
angstrom units	0.1000	millimicrons
angstrom units	0.1000	nanometres
astronomical units	1.4960×10^{8}	kilometres
astronomical units	** 1.4960×10^{11}	metres
bolts (US cloth)	* 120.0	linear feet
bolts (US cloth)	36.5760	metres
cables (US)	* 120.0	fathoms
cables (US)	720.0	feet
cables (US)	219.4560	metres
centimetres	1.0×10^{8}	angstrom units
centimetres	0.032 808 399	feet
centimetres	0.032 808 333	feet (US Survey)
centimetres	0.098 425 197	hands
centimetres	0.393 700 787	inches
centimetres	1.0×10^{-5}	kilometres
centimetres	0.032 808 399	links (engineers')
centimetres	0.049 709 695	links (US Survey)
centimetres	0.010	metres
centimetres	10 000.0	microns
centimetres	5.3996×10^{-6}	miles (nautical, Int)
centimetres	5.3961×10^{-6}	miles (nautical, UK)
centimetres	6.2137×10^{-6}	miles (statute)
centimetres	10.0	millimetres
centimetres	1.0×10^{7}	millimicrons
centimetres	393.700 787 4	mils
centimetres	2.371 063 016	picas (printers')
centimetres	28.452 756 19	points (printers')
centimetres	0.001 988 388	rods (surveyors')
centimetres	0.010 936 133	yards
chains (engineers')	3048.0	centimetres
chains (engineers')	1.515 151 515	chains (US Survey)
chains (engineers')	* 100.0	feet
chains (engineers')	99.999 80	feet (US Survey)
chains (engineers')	1200.0	inches
chains (engineers')	100.0	links (engineers')
chains (engineers')	151.515 151 5	links (US Survey)
chains (engineers')	30.480	metres
chains (engineers')	6.060 606 061	rods (surveyors')
chains (engineers')	33.333 333 333	yards

Linear Measure (Distance or depth)

MULTIPLY	BY	TO OBTAIN
chains (US Survey)	2011.680	centimetres
chains (US Survey)	0.660	chains (engineers')
chains (US Survey)	* 66.0	feet
chains (US Survey)	65.999 868 00	feet (US Survey)
chains (US Survey)	0.10	furlongs
chains (US Survey)	792.0	inches
chains (US Survey)	66.0	links (engineers')
chains (US Survey)	100.0	links (US Survey)
chains (US Survey)	20.116 80	metres
chains (US Survey)	0.012 50	miles (statute)
chains (US Survey)	4.0	rods (surveyors')
chains (US Survey)	22.0	yards
cubits	45.72	centimetres
cubits	* 1.5	feet
cubits	18.0	inches
cubits	0.457 20	metres
decimetres	10.0	centimetres
decimetres	0.328 083 990	feet
decimetres	0.328 083 333	feet (US Survey)
decimetres	3.937 007 874	inches
decimetres	0.10	metres
dekametres	1000.0	centimetres
dekametres	32.808 398 95	feet
dekametres	32.808 333 33	feet (US Survey)
dekametres	393.700 787 4	inches
dekametres	0.010	kilometres
dekametres	* 10.0	metres
dekametres	10.936 132 98	yards
digits	0.750	inches
ells	114.30	centimetres
ells	* 45.0	inches
ems (printers')	0.423 333 333 3	centimetres
ems (printers')	* 1/6	inches
ems (printers')	1.0	picas (printers')
fathoms	182.880	centimetres
fathoms	* 6.0	feet
fathoms	72.0	inches
fathoms	1.828 80	metres
fathoms	0.000 987 473	miles (nautical, Int)
fathoms	0.000 986 842	miles (nautical, UK)
fathoms	0.001 136 364	miles (statute)

47

Linear Measure (Distance or depth)

MULTIPLY	BY	TO OBTAIN
fathoms	2.0	yards
feet	30.480	centimetres
feet	0.015 151 515	chains (US Survey)
feet	1/6	fathoms
feet	0.999 998 000	feet (US Survey)
feet	1/660	furlongs
feet	* 12.0	inches
feet	0.000 304 80	kilometres
feet	* 0.304 80	metres
feet	304 800.0	microns
feet	0.000 164 579	miles (nautical, Int)
feet	1/6080	miles (nautical, UK)
feet	1/5280	miles (statute)
feet	304.80	millimetres
feet	1200.0	mils
feet	0.060 606 061	rods (surveyors')
feet	0.050	ropes (UK)
feet	1/3	yards
feet (US Survey)	30.480 060 96	centimetres
feet (US Survey)	0.010 000 020	chains (engineers')
feet (US Survey)	0.015 151 545	chains (US Survey)
feet (US Survey)	** 1.000 002 0	feet
feet (US Survey)	12.000 024 00	inches
feet (US Survey)	1.000 002 0	links (engineers')
feet (US Survey)	1.515 154 546	links (US Survey)
feet (US Survey)	0.340 800 610	metres
feet (US Survey)	0.000 189 394	miles (statute)
feet (US Survey)	0.060 606 182	rods (surveyors')
feet (US Survey)	0.333 334 0	yards
furlongs	20 116.80	centimetres
furlongs	6.60	chains (engineers')
furlongs	10.0	chains (US Survey)
furlongs	* 660.0	feet
furlongs	7920.0	inches
furlongs	0.201 168 0	kilometres
furlongs	201.1680	metres
furlongs	0.108 622 030	miles (nautical, Int)
furlongs	0.108 552 632	miles (nautical, UK)
furlongs	0.125	miles (statute)
furlongs	40.0	rods (surveyors')
furlongs	220.0	yards

Linear Measure (Distance or depth)

MULTIPLY	BY	TO OBTAIN
hand spans	22.860	centimetres
hand spans	0.1250	fathoms
hand spans	0.750	feet
hand spans	* 9.0	inches
hand spans	1.0	linear quarters (UK)
hands	10.160	centimetres
hands	* 4.0	inches
hectometres	10 000.0	centimetres
hectometres	1000.0	decimetres
hectometres	10.0	dekametres
hectometres	328.083 989 5	feet
hectometres	100.0	metres
hectometres	19.883 878 15	rods (surveyors')
hectometres	109.361 329 8	yards
inches	2.540×10^8	angstrom units
inches	2.540	centimetres
inches	0.001 262 626	chains (US Survey)
inches	1/18	cubits
inches	1/72	fathoms
inches	1/12	feet
inches	* 48.0	irons (footware)
inches	2.540×10^{-5}	kilometres
inches	* 12.0	lines
inches	1/12	links (engineers')
inches	0.126 262 626	links (US Survey)
inches	* 0.025 40	metres
inches	1.3715×10^{-5}	miles (nautical, Int)
inches	1.3706×10^{-5}	miles (nautical, UK)
inches	1.5783×10^{-5}	miles (statute)
inches	25.40	millimetres
inches	* 1000.0	mils
inches	6.022 500 060	picas (printers')
inches	** 72.270	points (printers')
inches	1/36	yards
irons (footware)	1/48	inches
irons (footware)	0.529 166 667	millimetres
kilometres	6.6850×10^{-9}	astronomical units
kilometres	1.0×10^5	centimetres
kilometres	3280.839 895	feet
kilometres	3280.833 333	feet (US Survey)
kilometres	39 370.078 74	inches

Linear Measure (Distance or depth)

MULTIPLY	BY	TO OBTAIN
kilometres	1.057×10^{-13}	light years
kilometres	1000.0	metres
kilometres	0.539 956 804	miles (nautical, Int)
kilometres	0.539 611 825	miles (nautical, UK)
kilometres	0.621 371 192	miles (statute)
kilometres	1.0×10^{6}	millimetres
kilometres	198.838 781 5	rods (surveyors')
kilometres	1093.613 298	yards
leagues (nautical, Int)	3038.0577	fathoms
leagues (nautical, Int)	18 228.346 46	feet
leagues (nautical, Int)	* 5.5560	kilometres
leagues (nautical, Int)	1.150 779 448	leagues (statute)
leagues (nautical, Int)	3.452 338 345	miles (statute)
leagues (nautical, UK)	* 18 240.0	feet
leagues (nautical, UK)	5.559 552 0	kilometres
leagues (nautical, UK)	1.000 639 309	leagues (nautical, Int)
leagues (nautical, UK)	1.151 515 152	leagues (statute)
leagues (nautical, UK)	3.454 545 455	miles (statute)
leagues (statute)	2640.0	fathoms
leagues (statute)	15 840.0	feet
leagues (statute)	4.828 032 0	kilometres
leagues (statute)	0.868 976 242	leagues (nautical, Int)
leagues (statute)	2.606 928 726	miles (nautical, Int)
leagues (statute)	* 3.0	miles (statute)
leagues (statute)	5280.0	yards
light years	63 278.636 36	astronomical units
light years	** 9.4605×10^{12}	kilometres
light years	5.8785×10^{12}	miles (statute)
lignes (watchmakers')	2.2560	millimetres
lines	0.211 666 667	centimetres
lines	1/12	inches
lines	2.116 666 667	millimetres
links (engineers')	30.480	centimetres
links (engineers')	0.010	chains (engineers')
links (engineers')	* 1.0	feet
links (engineers')	12.0	inches
links (engineers')	0.304 80	metres
links (US Survey)	0.010	chains (US Survey)
links (US Survey)	* 0.660	feet
links (US Survey)	0.659 998 680	feet (US Survey)
links (US Survey)	7.92	inches

Linear Measure (Distance or depth)

MULTIPLY	BY	TO OBTAIN
inks (US Survey)	0.201 168 0	metres
inks (US Survey)	0.000 125 0	miles (statute)
inks (US Survey)	0.040	rods (surveyors')
metres	1.0×10^{10}	angstrom units
metres	100.0	centimetres
metres	0.032 808 4	chains (engineers')
metres	0.049 709 7	chains (US Survey)
metres	0.546 806 7	fathoms
metres	3.280 839 895	feet
metres	3.280 833	feet (US Survey)
metres	0.004 970 97	furlongs
metres	39.370 078 74	inches
metres	0.001	kilometres
metres	3.280 84	links (engineers')
metres	4.970 97	links (US Survey)
metres	1.0×10^{-6}	megametres
metres	1.0×10^{6}	microns (μ)
metres	1/1852	miles (nautical, Int)
metres	0.000 539 612	miles (nautical, Int)
metres	0.000 621 371	miles (statute)
metres	1000.0	millimetres
metres	1.0×10^{9}	millimicrons
metres	39 370.078 74	mils
metres	0.198 838 782	rods (surveyors')
metres	1.093 613 298	yards
micromicrons	0.010	angstrom units
micromicrons	1.0×10^{-10}	centimetres
micromicrons	3.937×10^{-11}	inches
micromicrons	1.0×10^{-12}	metres
micromicrons	1.0×10^{-6}	microns
microns (μ)	* 10 000.0	angstrom units
microns (μ)	0.0001	centimetres
microns (μ)	3.2808×10^{-6}	feet
microns (μ)	3.9370×10^{-5}	inches
microns (μ)	* 1.0×10^{-6}	metres
microns (μ)	39.370 078 74	microinches
microns (μ)	0.0010	millimetres
microns (μ)	1000.0	millimicrons
microns (μ)	0.039 370 079	mils
miles (nautical, Int)	8.439 049 286	cables (US)
miles (nautical, Int)	185 200.0	centimetres

Linear Measure (Distance or depth)

MULTIPLY	BY	TO OBTAIN
miles (nautical, Int)	1012.685 914	fathoms
miles (nautical, Int)	6076.115 486	feet
miles (nautical, Int)	6076.103 334	feet (US Survey)
miles (nautical, Int)	72 913.385 83	inches
miles (nautical, Int)	1.8520	kilometres
miles (nautical, Int)	1/3	leagues (nautical, Int)
miles (nautical, Int)	* 1852.0	metres
miles (nautical, Int)	0.999 361 10	miles (nautical, UK)
miles (nautical, Int)	1.150 779 448	miles (statute)
miles (nautical, Int)	1.8520×10^6	millimetres
miles (nautical, Int)	2025.371 829	yards
miles (nautical, UK)	8.444 444 444	cables (US)
miles (nautical, UK)	185 318.40	centimetres
miles (nautical, UK)	1013.333 333	fathoms
miles (nautical, UK)	* 6080.0	feet
miles (nautical, UK)	72 960.0	inches
miles (nautical, UK)	1.853 184 0	kilometres
miles (nautical, UK)	1853.184 000	metres
miles (nautical, UK)	1.000 639 309	miles (nautical, Int)
miles (nautical, UK)	1.151 515 152	miles (statute)
miles (nautical, UK)	$1.853\ 18 \times 10^6$	millimetres
miles (nautical, UK)	2026.666 667	yards
miles (statute)	160 934.40	centimetres
miles (statute)	52.80	chains (engineers')
miles (statute)	80.0	chains (US Survey)
miles (statute)	* 5280.0	feet
miles (statute)	5279.9894	feet (US Survey)
miles (statute)	* 8.0	furlongs
miles (statute)	63 360.0	inches
miles (statute)	1.609 344 0	kilometres
miles (statute)	1.057×10^{-13}	light years
miles (statute)	8000.0	links (US Survey)
miles (statute)	1609.3440	metres
miles (statute)	0.868 976 242	miles (nautical, Int)
miles (statute)	0.868 421 053	miles (nautical, UK)
miles (statute)	$1.609\ 34 \times 10^6$	millimetres
miles (statute)	320.0	rods (surveyors')
miles (statute)	1760.0	yards
millimetres	1.0×10^7	angstrom units
millimetres	0.10	centimetres
millimetres	0.010	decimetres

Linear Measure (Distance or depth)

MULTIPLY	BY	TO OBTAIN
millimetres	0.000 10	dekametres
millimetres	0.003 280 840	feet
millimetres	0.039 370 079	inches
millimetres	1.0×10^{-6}	kilometres
millimetres	0.001	metres
millimetres	1000.0	microns (μ)
millimetres	39.370 078 74	mils
millimetres	0.001 093 613	yards
millimicrons	10.0	angstrom units
millimicrons	1.0×10^{-7}	centimetres
millimicrons	3.9370×10^{-8}	inches
millimicrons	1.0×10^{-9}	metres
millimicrons	0.0010	microns
millimicrons	1.0×10^{-6}	millimetres
mils	0.002 54	centimetres
mils	8.333×10^{-5}	feet
mils	0.001	inches
mils	2.54×10^{-8}	kilometres
mils	25.4	microns (μ)
mils	0.0254	millimetres
mils	2.7778×10^{-5}	yards
myriametres (US)	* 10.0	kilometres
myriametres (US)	10 000.0	metres
paces	76.20	centimetres
paces	0.0250	chains (engineers')
paces	0.037 878 788	chains (US Survey)
paces	* 2.50	feet
paces	7.5	hands
paces	30.0	inches
paces	0.125	ropes (UK)
palms	7.620	centimetres
palms	0.002 50	chains (engineers')
palms	1/6	cubits
palms	0.250	feet
palms	0.750	hands
palms	* 3.0	inches
parsecs	206 265.5840	astronomical units
parsecs	3.0857×10^{13}	kilometres
parsecs	** 3.261 60	light years
parsecs	1.9174×10^{13}	miles (statute)
perches	16.50	feet

53

Linear Measure (Distance or depth)

MULTIPLY	BY		TO OBTAIN
perches		5.029 20	metres
perches		1.0	poles
perches		1.0	rods (surveyors')
perches	*	5.50	yards
picas (printers')		0.423 333 333	centimetres
picas (printers')		1.0	ems
picas (printers')		1/6	inches
picas (printers')	*	12.0	points (printers')
pica ems (UK, US)	**	0.1660	inches
points (Didot)	*	0.3760	millimetres
points (UK, US)	**	0.3550	millimetres
points (printers')		0.035 145 90	centimetres
points (printers')		0.013 837 0	inches
points (printers')		1/12	picas (printers')
poles		5.029 20	metres
poles	*	1.0	perches
poles	*	1.0	rods (surveyors')
poles	*	5.5	yards
rods (surveyors')		502.920	centimetres
rods (surveyors')		0.1650	chains (engineers')
rods (surveyors')		0.250	chains (US Survey)
rods (surveyors')		16.50	feet
rods (surveyors')		16.499 967 0	feet (US Survey)
rods (surveyors')		0.0250	furlongs
rods (surveyors')		198.0	inches
rods (surveyors')		16.5	links (engineers')
rods (surveyors')	*	25.0	links (US Survey)
rods (surveyors')		5.029 20	metres
rods (surveyors')		0.003 125 0	miles (statute)
rods (surveyors')		1.0	perches
rods (surveyors')		1.0	poles
rods (surveyors')	*	5.50	yards
ropes (UK)	*	20.0	feet
ropes (UK)		6.0960	metres
ropes (UK)		6.666 667	yards
thou (mils)		2.540×10^{-5}	metres
thou (mils)	*	0.025 40	millimetres
yards		91.440	centimetres
yards		0.030	chains (engineers')
yards		0.045 454 545	chains (US Survey)
yards		2.0	cubits

Linear Measure (Distance or depth)

MULTIPLY	BY	TO OBTAIN
yards	0.50	fathoms
yards	* 3.0	feet
yards	2.999 994 0	feet (US Survey)
yards	0.004 545 455	furlongs
yards	4.0	hand spans
yards	36.0	inches
yards	0.000 914 40	kilometres
yards	4.0	linear quarters (UK)
yards	0.914 40	metres
yards	0.000 493 737	miles (nautical, Int)
yards	0.000 493 421	miles (nautical, UK)
yards	0.000 568 182	miles (statute)
yards	914.40	millimetres
yards	0.181 818 182	perches
yards	0.181 818 182	poles (UK)
yards	0.181 818 182	rods (surveyors')
yarns/hanks of cotton	2520.0	feet
yarns/hanks of cotton	768.0960	metres
yarns/hanks of cotton	* 840.0	yards
yarns/hanks of wool	1680.0	feet
yarns/hanks of wool	109.7280	metres
yarns/hanks of wool	* 560.0	yards
yarns/leas of linen	900.0	feet
yarns/leas of linen	274.320	metres
yarns/leas of linen	* 300.0	yards
yarns/skeins of wool	768.0	feet
yarns/skeins of wool	234.086 40	metres
yarns/skeins of wool	* 256.0	yards
yarns/spindles of jute	43 200.0	feet
yarns/spindles of jute	13 167.360	metres
yarns/spindles of jute	* 14 400.0	yards

MASS
Dimensions: M

MULTIPLY	BY	TO OBTAIN
atomic mass units (chem)	1.660×10^{-24}	grams
atomic mass units (phys)	1.659×10^{-24}	grams
bags of cement	94.0	pounds of cement
barrels of cement	376.0	pounds of cement
carats (diamond)	200.0	milligrams

Mass

MULTIPLY	BY	TO OBTAIN
carats (gold)	4.166 666 667	% purity of gold
carats (metric)	2.539 725 235	grains
carats (metric)	0.20	grams
carats (metric)	* 200.0	milligrams
carats (gold parts/24)	41.666 666 67	milligrams/gram
centigrams	0.154 323 582	grains
centigrams	0.010	grams
centigrams	1.0×10^{-5}	kilograms
daltons (chem)	166.024×10^{-26}	grams
daltons (phys)	165.979×10^{-26}	grams
decigrams	0.10	grams
dekagrams	10.0	grams
grams	5.0	carats (metric)
grams	10.0	decigrams
grams	0.10	dekagrams
grams	0.257 205 970	drams (apothecary/troy)
grams	0.564 383 386	drams (avoirdupois)
grams	980.6650	dynes
grams	15.432 358 20	grains
grams	9.8067×10^{-5}	joules/centimetre
grams	0.009 806 650	joules/metre (newtons)
grams	0.0010	kilograms
grams	1.0×10^{6}	micrograms
grams	1000.0	milligrams
grams	0.035 273 962	ounces (avoirdupois)
grams	0.032 150 746	ounces (apothecary/troy)
grams	0.070 931 635	poundals
grams	0.002 204 623	pounds (avoirdupois)
grams	0.002 679 229	pounds (apothecary/troy)
grams	0.771 617 910	scruples (apothecary)
grams	1.0×10^{-6}	tonnes (metric)
hectograms	100.0	grams
hectograms	7.093 163 459	poundals
hectograms	0.267 922 885	pounds (apothecary/troy)
hectograms	0.220 462 260	pounds (avoirdupois)
kilograms	257.205 970	drams (apothecary/troy)
kilograms	564.383 385 6	drams (avoirdupois)
kilograms	980 665.0	dynes
kilograms	15 432.358 20	grains
kilograms	1000.0	grams
kilograms	0.019 684 130	hundredweights (long)

Mass

MULTIPLY	BY	TO OBTAIN
kilograms	0.022 046 226	hundredweights (short)
kilograms	0.098 066 50	joules/centimetre
kilograms	32.150 746 25	ounces (apothecary/troy)
kilograms	35.273 961 60	ounces (avoirdupois)
kilograms	643.014 925 0	pennyweights
kilograms	70.931 634 59	poundals
kilograms	2.204 622 60	pounds (avoirdupois)
kilograms	2.679 228 854	pounds (apothecary/troy)
kilograms	771.617 910	scruples (apothecary)
kilograms	0.068 521 765	slugs
kilograms	0.0010	tonnes (metric)
kilograms	0.000 984 207	tons (long)
kilograms	0.001 102 311	tons (short)
micrograms	1.0×10^{-6}	grams
micrograms	0.0010	milligrams
milliers	1000.0	kilograms
milligrams	0.0050	carats (metric)
milligrams	0.000 257 206	drams (apothecary/troy)
milligrams	0.000 564 383	drams (avoirdupois)
milligrams	0.015 432 358	grains
milligrams	0.0010	grams
milligrams	1.0000×10^{-6}	kilograms
milligrams	3.2151×10^{-5}	ounces (apothecary/troy)
milligrams	3.5274×10^{-5}	ounces (avoirdupois)
milligrams	0.000 643 015	pennyweights
milligrams	$2.6792/10^{-6}$	pounds (apothecary/troy)
milligrams	2.2046×10^{-6}	pounds (avoirdupois)
milligrams	0.000 771 618	scruples (apothecary)
myriagrams	10.0	kilograms
pounds (avoirdupois)	** 453.592 374 5	grams
pounds (avoirdupois)	0.453 592 375	kilograms
pounds water (15.18 °C)	0.016 036 761	cubic feet
pounds water (15.18 °C)	27.711 523 82	cubic inches
pounds water (15.18 °C)	0.000 454 111	cubic metres
pounds water (15.18 °C)	0.099 890 349	gallons (UK)
pounds water (15.18 °C)	0.119 963 307	gallons (US liquid)
pounds water (17.0 °C)	0.016 039 608	cubic feet
pounds water (17.0 °C)	27.716 442 47	cubic inches
pounds water (17.0 °C)	0.000 454 191	cubic metres
pounds water (17.0 °C)	0.099 908 078	gallons (UK)
pounds water (17.0 °C)	0.119 984 599	gallons (US liquid)

Mass

MULTIPLY	BY	TO OBTAIN
pounds water (4.0 °C)	0.016 019 425	cubic feet
pounds water (4.0 °C)	27.681 565 78	cubic inches
pounds water (4.0 °C)	0.000 453 620	cubic metres
pounds water (4.0 °C)	0.099 782 360	gallons (UK)
pounds water (4.0 °C)	0.119 833 618	gallons (US liquid)
slugs	14.593 903 08	kilograms
slugs	32.174 048 56	pounds
tonnes (metric)	1.0×10^6	grams
tonnes (metric)	1000.0	kilograms
tonnes (metric)	2204.622 60	pounds
tonnes (metric)	0.984 206 518	tons (long)
tonnes (metric)	1.102 311 30	tons (short)
tons (long)	1016.046 919	kilograms
tons (long)	2240.0	pounds (avoirdupois)
tons (long)	1.016 046 919	tonnes (metric)
tons (long)	1.120	tons (short)
tons (short)	907.184 749 0	kilograms
tons (short)	29 166.666 67	ounces (apothecary/troy)
tons (short)	32 000.0	ounces (avoirdupois)
tons (short)	2430.555 56	pounds (apothecary/troy)
tons (short)	2000.0	pounds (avoirdupois)
tons (short)	0.907 184 749	tonnes (metric)
tons (short)	0.892 857 143	tons (long)

MASS FLOW
Dimensions: M/T

MULTIPLY	BY	TO OBTAIN
kilograms/second	3600.0	kilograms/hour
pounds/hour	0.125 997 882	grams/second
pounds/hour	0.453 592 375	kilograms/hour
pounds/hour	0.000 125 998	kilograms/second
pounds/minute	27.215 542 47	kilograms/hour
pounds/minute	0.453 592 375	kilograms/minute
tonnes of water/24 hours	1.473 125 282	cubic feet/hour
tonnes of water/24 hours	0.183 662 373	gallons (US)/minute
tonnes of water/24 hours	91.859 275 0	pounds of water/hour
tonnes(metric)/24 hours	1000.0	kilograms/hour
tonnes(metric)/24 hours	0.011 574 074	kilograms/second
tonnes(metric)/24 hours	0.277 777 778	kilograms/second
tons(long) water/24 h	1.496 764 405	cubic feet/hour

Mass Flow

MULTIPLY	BY	TO OBTAIN
tons(long) water/24 h	0.155 384 987	gallons (UK)/minute
tons(long) water/24 h	0.186 609 588	gallons (US)/minute
tons(long)/24 hours	0.011 759 802	kilograms/second
tons(long)/24 hours	1.016 046 919	tonnes(metric)/24 hours
tons(long)/hour	1016.046 919	kilograms/hour
tons(long)/hour	0.282 235 255	kilograms/second
tons(short) water/24 h	1.336 396 790	cubic feet/hour
tons(short) water/24 h	0.166 615 704	gallons (US)/minute
tons(short)/24 hours	0.010 499 823	kilograms/second
tons(short)/24 hours	0.907 184 749	tonnes(metric)/24 hours
tons(short)/hour	907.184 749 0	kilograms/hour
tons(short)/hour	0.251 995 764	kilograms/second

MASS FLUX DENSITY
Dimensions: M/TL^2

MULTIPLY	BY	TO OBTAIN
grams/second cm^2	10.0	kg/second metre^2
kg/hour foot^2	0.002 989 975	kg/second metre^2
kg/hour metre^2	1/3600	kg/second metre^2
pounds/hour foot^2	1.356 229 913	grams/second metre^2
pounds/hour foot^2	0.001 356 230	kg/second metre^2
pounds/second foot^2	4.882 427 687	kg/second metre^2

MOMENT OF INERTIA
Dimensions: ML^2

MULTIPLY	BY	TO OBTAIN
gram centimetre^2	0.0010	kg centimetre^2
gram centimetre^2	1.0197×10^{-6}	kgf centimetre second^2
gram centimetre^2	1.0197×10^{-8}	kgf metre second^2
gram centimetre^2	1.0×10^{-7}	kilogram metre^2
gram centimetre^2	0.000 014 161	ounce inch second^2
gram centimetre^2	0.005 467 475	ounce inch^2
gram centimetre^2	7.3756×10^{-8}	pound foot second^2
gram centimetre^2	2.3730×10^{-6}	pound foot^2
gram centimetre^2	8.8508×10^{-7}	pound inch second^2
kg centimetre^2	0.001 019 716	kgf centimetre second^2
kg centimetre^2	0.000 010 197	kgf metre second^2
kg centimetre^2	0.000 10	kilogram metre^2
kg centimetre^2	0.014 161 193	ounce inch second^2
kg centimetre^2	5.467 474 983	ounce inch^2

Moment of Inertia

MULTIPLY	BY	TO OBTAIN
kg centimetre^2	0.000 073 756	pound foot second^2
kg centimetre^2	0.002 373 036	pound foot^2
kg centimetre^2	0.000 885 075	pound inch second^2
kg centimetre^2	0.341 717 186	pound inch^2
kgf centimetre second^2	980.6650	kg centimetre^2
kgf centimetre second^2	0.010	kgf metre second^2
kgf centimetre second^2	0.098 066 50	kilogram metre^2
kgf centimetre second^2	13.887 386 46	ounce inch second^2
kgf centimetre second^2	5361.761 354	ounce inch^2
kgf centimetre second^2	0.072 330 138	pound foot second^2
kgf centimetre second^2	2.327 153 366	pound foot^2
kgf centimetre second^2	0.867 961 654	pound inch second^2
kgf centimetre second^2	335.110 084 6	pound inch^2
kgf metre second^2	98 066.50	kg centimetre^2
kgf metre second^2	100.0	kgf centimetre second^2
kgf metre second^2	9.806 650	kilogram metre^2
kgf metre second^2	1388.738 646	ounce inch second^2
kgf metre second^2	536 176.1354	ounce inch^2
kgf metre second^2	7.233 013 779	pound foot second^2
kgf metre second^2	232.715 336 6	pound foot^2
kgf metre second^2	86.796 165 35	pound inch second^2
kgf metre second^2	33 511.008 46	pound inch^2
kgf metre second^2	7.233 013 780	slug foot^2
kilogram metre^2	10 000.0	kg centimetre^2
kilogram metre^2	10.197 162 13	kgf centimetre second^2
kilogram metre^2	0.101 971 621	kgf metre second^2
kilogram metre^2	141.611 931 3	ounce inch^2
kilogram metre^2	54 674.749 83	ounce inch^2
kilogram metre^2	0.737 562 142	pound foot second^2
kilogram metre^2	23.730 360 18	pound foot^2
kilogram metre^2	8.850 745 705	pound inch second^2
kilogram metre^2	3417.171 866	pound inch^2
ounce inch second^2	70.615 518 85	kg centimetre^2
ounce inch second^2	0.072 007 789	kgf centimetre second^2
ounce inch second^2	0.000 720 078	kgf metre second^2
ounce inch second^2	0.007 061 552	kilogram metre^2
ounce inch second^2	386.088 582 8	ounce inch^2
ounce inch second^2	0.005 208 333	pound foot second^2
ounce inch second^2	0.167 573 170	pound foot^2
ounce inch second^2	0.062 50	pound inch second^2
ounce inch second^2	24.130 536 42	pound inch^2

Moment of Inertia

MULTIPLY	BY	TO OBTAIN
ounce inch²	0.182 899 785	kg centimetre²
ounce inch²	0.000 186 506	kgf centimetre second²
ounce inch²	1.8651×10^{-6}	kgf metre second²
ounce inch²	1.8290×10^{-5}	kilogram metre²
ounce inch²	0.002 590 079	ounce inch second²
ounce inch²	1.3490×10^{-5}	pound foot second²
ounce inch²	0.000 434 028	pound foot²
ounce inch²	0.000 161 880	pound inch second²
ounce inch²	0.062 50	pound inch²
pound foot second²	13 558.179 61	kg centimetre²
pound foot second²	13.825 495 57	kgf centimetre second²
pound foot second²	0.138 254 956	kgf metre second²
pound foot second²	1.355 817 961	kilogram metre²
pound foot second²	192.0	ounce inch second²
pound foot second²	74 129.007 83	ounce inch²
pound foot second²	32.174 048 54	pound foot²
pound foot second²	12.0	pound inch second²
pound foot second²	4633.062 989	pound inch²
pound foot²	421.401 105 0	kg centimetre²
pound foot²	0.429 709 539	kgf centimetre second²
pound foot²	0.004 297 095	kgf metre second²
pound foot²	0.042 140 111	kilogram metre²
pound foot²	5.967 542 429	ounce inch second²
pound foot²	2304.0	ounce inch²
pound foot²	0.031 080 950	pound foot second²
pound foot²	0.372 971 402	pound inch second²
pound foot²	144.0	pound inch²
pound inch second²	1129.848 301	kg centimetre²
pound inch second²	1.152 124 631	kgf centimetre second²
pound inch second²	0.011 521 246	kgf metre second²
pound inch second²	0.112 984 830	kilogram metre²
pound inch second²	16.0	ounce inch second²
pound inch second²	6177.417 320	ounce inch²
pound inch second²	1/12	pound foot second²
pound inch second²	2.681 170 713	pound foot²
pound inch second²	386.088 582 5	pound inch²
pound inch²	2.926 396 563	kg centimetre²
pound inch²	0.002 984 094	kgf centimetre second²
pound inch²	2.9841×10^{-5}	kgf metre second²
pound inch²	0.000 292 640	kilogram metre²
pound inch²	0.041 441 267	ounce inch second²

Moment of Inertia

MULTIPLY	BY	TO OBTAIN
pound inch^2	16.0	ounce inch^2
pound inch^2	0.000 215 840	pound foot second^2
pound inch^2	1/144	pound foot^2
pound inch^2	0.002 590 079	pound inch second^2
slug foot^2	13 558.179 61	kg centimetre^2
slug foot^2	13.825 495 57	kgf centimetre second^2
slug foot^2	0.138 254 956	kgf metre second^2
slug foot^2	1.355 817 961	kilogram metre^2
slug foot^2	192.0	ounce inch second^2
slug foot^2	74 129.007 83	ounce inch^2
slug foot^2	1.0	pound foot second^2
slug foot^2	32.174 048 56	pound foot^2
slug foot^2	12.0	pound inch second^2
slug foot^2	4633.062 989	pound inch^2

PAPER
Dimensions: None

MULTIPLY	BY	TO OBTAIN
quires	25.0	sheets
reams	20.0	quires
reams	480.0	sheets
reams (mean)	500.0	sheets
reams (printers')	516.0	sheets

POWER
Dimensions: ML^2/T^3

MULTIPLY	BY	TO OBTAIN
Btu/hour	2.930 71 × 10^6	ergs/second
Btu/hour	778.169 270 1	foot pounds/hour
Btu/hour	0.216 158 131	foot pounds/second
Btu/hour	0.069 998 823	gram calories/second
Btu/hour	0.000 398 466	horsepower (metric)
Btu/hour	0.000 393 015	horsepower (UK)
Btu/hour	0.252 083 828	kilogram calories/hour
Btu/hour	0.000 293 071	kilowatts
Btu/hour	0.293 071 073	watts (joules/second)
Btu/minute	1.758 43 × 10^8	ergs/second
Btu/minute	778.169 270 1	foot pounds/minute
Btu/minute	12.969 487 84	foot pounds/second
Btu/minute	0.023 580 887	horsepower (UK)

Power

MULTIPLY	BY	TO OBTAIN
Btu/minute	0.251 995 764	kilogram calories/min
Btu/minute	107.585 756 9	kgf metres/minute
Btu/minute	0.017 584 264	kilowatts
Btu/minute	0.417 844 312	lb of ice melted/hour
Btu/minute	17.584 264 38	watts (joules/second)
Btu/second	3600.0	Btu/hour
Btu/second	60.0	Btu/minute
Btu/second	1.434 476 759	cheval-vapeur
Btu/second	1.0551×10^{10}	ergs/second
Btu/second	778.169 270 1	foot pounds/second
Btu/second	1.434 476 759	horsepower (metric)
Btu/second	1.414 853 218	horsepower (UK)
Btu/second	907.184 749 0	kilogram calories/hour
Btu/second	15.185 029 69	kilogram calories/min
Btu/second	107.585 756 9	kgf metres/second
Btu/second	1.055 055 863	kilowatts
Btu/second	1055.055 863	watts (joules/second)
cheval-vapeur	* 1.0	horsepower (metric)
cheval-vapeur	0.986 320 071	horsepower (UK)
cheval-vapeur	75.0	kgf metres/second
cheval-vapeur	735.498 750	watts (joules/second)
Chu/hour	0.000 527 528	kilowatts
Chu/hour	0.527 527 931	watts
cubic metre atm/hour	0.028 145 833	kilowatts
cubic metre atm/hour	28.145 833 33	watts (joules/second)
ergs/second	5.6869×10^{-9}	Btu/minute
ergs/second	1.0	dyne centimetres/second
ergs/second	4.4254×10^{-6}	foot pounds/minute
ergs/second	7.3756×10^{-8}	foot pounds/second
ergs/second	1.4331×10^{-6}	gram calories/minute
ergs/second	0.001 019 716	gram(f) cm/second
ergs/second	1.360×10^{-10}	horsepower (metric)
ergs/second	1.341×10^{-10}	horsepower (UK)
ergs/second	1.4331×10^{-9}	kg calories/minute
ergs/second	1.0×10^{-10}	kilowatts
ergs/second	1.0×10^{-7}	watts (joules/second)
oot poundals/hour	1.1706×10^{-8}	kilowatts
oot poundals/hour	1.1706×10^{-5}	watts
oot poundals/minute	0.002 528 407	kilowatts
oot poundals/minute	2.528 406 605	watts
oot poundals/second	4.2140×10^{-5}	kilowatts

Power

MULTIPLY	BY	TO OBTAIN
foot poundals/second	0.042 140 110	watts
foot pounds/hour	2.1418×10^{-5}	Btu/minute
foot pounds/hour	3.5696×10^{-7}	Btu/second
foot pounds/hour	$2.259\ 70 \times 10^{5}$	ergs/minute
foot pounds/hour	1/60	foot pounds/minute
foot pounds/hour	0.005 397 193	gram calories/minute
foot pounds/hour	5.1206×10^{-7}	horsepower (metric)
foot pounds/hour	5.0505×10^{-7}	horsepower (UK)
foot pounds/hour	3.7662×10^{-7}	kilowatts
foot pounds/hour	0.000 376 616	watts
foot pounds/minute	0.077 104 047	Btu/hour
foot pounds/minute	2.1418×10^{-5}	Btu/second
foot pounds/minute	0.042 835 582	Chu/hour
foot pounds/minute	1.1899×10^{-5}	Chu/second
foot pounds/minute	$8.134\ 91 \times 10^{8}$	ergs/hour
foot pounds/minute	$2.259\ 70 \times 10^{5}$	ergs/second
foot pounds/minute	60.0	foot pounds/hour
foot pounds/minute	1/60	foot pounds/second
foot pounds/minute	19.429 893 21	gram calories/hour
foot pounds/minute	0.005 397 193	gram calories/second
foot pounds/minute	3.0723×10^{-5}	horsepower (metric)
foot pounds/minute	3.0303×10^{-5}	horsepower (UK)
foot pounds/minute	0.000 323 832	kilogram calories/min
foot pounds/minute	0.138 254 954	kgf metres/minute
foot pounds/minute	2.2597×10^{-5}	kilowatts
foot pounds/minute	0.022 596 966	watts (joules/second)
foot pounds/second	4.626 242 820	Btu/hour
foot pounds/second	0.077 104 047	Btu/minute
foot pounds/second	0.001 285 067	Btu/second
foot pounds/second	2.570 134 901	Chu/hour
foot pounds/second	0.042 835 582	Chu/minute
foot pounds/second	4.8809×10^{10}	ergs/hour
foot pounds/second	$8.134\ 91 \times 10^{8}$	ergs/minute
foot pounds/second	$1.355\ 82 \times 10^{7}$	ergs/second
foot pounds/second	0.323 831 554	gram calories/second
foot pounds/second	13 825.495 43	gram(f) cm/second
foot pounds/second	0.001 843 399	horsepower (metric)
foot pounds/second	0.001 818 182	horsepower (UK)
foot pounds/second	1.165 793 593	kilogram calories/hour
foot pounds/second	0.019 429 893	kilogram calories/min
foot pounds/second	0.001 355 818	kilowatts

Power

MULTIPLY	BY	TO OBTAIN
foot pounds/second	1.355 817 948	watts (joules/second)
gram calories/hour	0.003 968 321	Btu/hour
gram calories/hour	11 630.0	ergs/second
gram calories/hour	0.001 163 0	watts
gram calories/second	14.285 954 45	Btu/hour
gram calories/second	0.004 186 80	kilowatts
gram calories/second	4.186 80	watts (joules/second)
gram(f) cm/second	9.2949×10^{-8}	Btu/second
gram(f) cm/second	980.6650	ergs/second
gram(f) cm/second	7.2330×10^{-5}	foot pounds/second
gram(f) cm/second	2.3423×10^{-5}	gram calories/second
gram(f) cm/second	1.3333×10^{-7}	horsepower (metric)
gram(f) cm/second	1.3151×10^{-7}	horsepower (UK)
gram(f) cm/second	9.8067×10^{-8}	kilowatts
gram(f) cm/second	9.8067×10^{-5}	watts (joules/second)
hectowatts	100.0	watts
horsepower (boiler)	33 472.119 54	Btu/hour
horsepower (boiler)	9.8097×10^{10}	ergs/second
horsepower (boiler)	434 116.2472	foot pounds/minute
horsepower (boiler)	140 580.5387	gram calories/minute
horsepower (boiler)	13.149 745 29	horsepower (electric)
horsepower (boiler)	13.337 493 76	horsepower (metric)
horsepower (boiler)	13.155 037 79	horsepower (UK)
horsepower (boiler)	9.809 709 990	kilowatts
horsepower (boiler)	* 34.50	lb water evap/h 100 °C
horsepower (boiler)	9809.709 990	watts (joules/second)
horsepower (electric)	2545.457 633	Btu/hour
horsepower (electric)	7.460×10^9	ergs/second
horsepower (electric)	33 013.281 81	foot pounds/minute
horsepower (electric)	550.221 363 5	foot pounds/second
horsepower (electric)	178.179 038 9	gram calories/second
horsepower (electric)	0.076 047 10	horsepower (boiler)
horsepower (electric)	1.014 277 727	horsepower (metric)
horsepower (electric)	1.000 402 479	horsepower (UK)
horsepower (electric)	* 746.0	joules/second (watts)
horsepower (electric)	641. 444 540 0	kilogram calories/hour
horsepower (electric)	0.7460	kilowatts
horsepower (electric)	17.726 749 86	lb of ice melted/hour
horsepower (metric)	2509.625 881	Btu/hour
horsepower (metric)	$7.354 99 \times 10^9$	ergs/second
horsepower (metric)	32 548.562 34	foot pounds/minute

Power

MULTIPLY	BY	TO OBTAIN
horsepower (metric)	542.476 039 0	foot pounds/second
horsepower (metric)	632 415.0902	gram calories/hour
horsepower (metric)	0.074 976 605	horsepower (boiler)
horsepower (metric)	0.985 923 257	horsepower (electric)
horsepower (metric)	0.986 320 071	horsepower (UK)
horsepower (metric)	* 75.0	kgf metres/second
horsepower (metric)	0.735 498 750	kilowatts
horsepower (metric)	17.477 214 97	lb of ice melted/hour
horsepower (metric)	735.498 750 0	watts (joules/second)
horsepower (UK)	2544.433 552	Btu/hour
horsepower (UK)	42.407 225 87	Btu/minute
horsepower (UK)	7.4570×10^9	ergs/second
horsepower (UK)	1.980×10^6	foot pounds/hour
horsepower (UK)	33 000.0	foot pounds/minute
horsepower (UK)	* 550.0	foot pounds/second
horsepower (UK)	641 186.4758	gram calories/hour
horsepower (UK)	0.076 016 505	horsepower (boiler)
horsepower (UK)	0.999 597 683	horsepower (electric)
horsepower (UK)	1.013 869 665	horsepower (metric)
horsepower (UK)	10.686 441 26	kilogram calories/min
horsepower (UK)	0.745 699 871	kilowatts
horsepower (UK)	17.719 618 09	lb of ice melted/hour
horsepower (UK)	0.212 033 288	tons of refrigeration
horsepower (UK)	745.699 871 4	watts (joules/second)
joules/second (watts)	0.056 869 027	Btu/minute
joules/second (watts)	1.0×10^7	dyne centimetres/second
joules/second (watts)	1.0×10^7	ergs/second
joules/second (watts)	0.737 562 150	foot pounds/second
joules/second (watts)	14.330 753 80	gram calories/minute
joules/second (watts)	10 197.162 13	gram(f) cm/second
joules/second (watts)	0.001 340 483	horsepower (electric)
joules/second (watts)	0.001 359 622	horsepower (metric)
joules/second (watts)	0.001 341 022	horsepower (UK)
joules/second (watts)	0.014 330 754	kg calories/minute
joules/second (watts)	0.0010	kilowatts
joules/second (watts)	0.023 762 399	lb of ice melted/hour
kilocalories/hour	0.001 163 0	kilowatts
kilocalories/hour	1.1630	watts (joules/second)
kilocalories/minute	51.467 086 78	foot pounds/second
kilocalories/minute	0.094 874 396	horsepower (metric)
kilocalories/minute	0.093 576 521	horsepower (UK)

Power

MULTIPLY	BY	TO OBTAIN
kilocalories/minute	0.069 780	kilowatts
kilocalories/second	4.186 80	kilowatts
kilocalories/second	4186.8	watts
kgf metres/second	0.013 333 333	horsepower (metric)
kgf metres/second	0.013 150 934	horsepower (UK)
kgf metres/second	0.009 806 650	kilowatts
kgf metres/second	9.806 650	watts (joules/second)
kilowatts	3412.141 60	Btu/hour
kilowatts	56.869 026 67	Btu/minute
kilowatts	1.0×10^{10}	ergs/second
kilowatts	$1.423\ 82 \times 10^{6}$	foot poundals/minute
kilowatts	$2.655\ 22 \times 10^{6}$	foot pounds/hour
kilowatts	$4.425\ 37 \times 10^{4}$	foot pounds/minute
kilowatts	737.562 149 5	foot pounds/second
kilowatts	$1.019\ 72 \times 10^{7}$	gram(f) cm/second
kilowatts	0.101 939 813	horsepower (boiler)
kilowatts	1.359 621 617	horsepower (metric)
kilowatts	1.341 022 090	horsepower (UK)
kilowatts	3.60×10^{6}	joules/hour
kilowatts	14.330 753 80	kilogram calories/min
kilowatts	$3.670\ 98 \times 10^{5}$	kgf metres/hour
kilowatts	0.284 341 323	tons of refrigeration
kilowatts	1000.0	watts (joules/second)
lb of ice melted/hour	42.083 292 53	watts (joules/second)
poncelots	4/3	horsepower (metric)
poncelots	1.315 093 428	horsepower (UK)
poncelots	100.0	kgf metres/second
poncelots	980.6650	watts (joules/second)
pounds feet/second	0.138 254 954	kgf metres/second
therms/hour	29.307 107 31	kilowatts
therms/hour	29 307.107 31	watts (joules/second)
tonne calories/hour	1.1630	kilowatts
tonne calories/hour	1163.0	watts (joules/second)
tons of refrigeration	288 003.8590	Btu/24 hours
tons of refrigeration	12 000.160 79	Btu/hour
tons of refrigeration	3.516 90	kilowatts
tons of refrigeration	** 3516.90	watts
watts	3.412 141 599	Btu/hour
watts	0.056 869 027	Btu/minute
watts	0.000 947 817	Btu/second
watts	1.0×10^{7}	ergs/second

Power

MULTIPLY	BY	TO OBTAIN
watts	44.253 728 97	foot pounds/minute
watts	0.737 562 150	foot pounds/second
watts	859.845 227 9	gram calories/hour
watts	14.330 753 80	gram calories/minute
watts	0.000 101 940	horsepower (boiler)
watts	0.001 359 622	horsepower (metric)
watts	0.001 341 022	horsepower (UK)
watts	1.0	joules/second
watts	0.014 330 754	kg calories/minute
watts	0.0010	kilowatts
watts	35.529 237 61	litre atmospheres/hour
watts	0.000 284 341	tons of refrigeration

PRESSURE or STRESS
Dimensions: M/LT^2

MULTIPLY	BY	TO OBTAIN
at (metric atmosphere)	* 1.0	kgf/square centimetre
at (metric atmosphere)	98.066 50	kilonewtons/sq metre
at (metric atmosphere)	1.0	kilopond(kp)/sq cm
at (metric atmosphere)	98 066.50	pascals
atmospheres (Standard)	* 1.013 250	bar
atmospheres (Standard)	75.999 989 18	cm of mercury (0 °C)
atmospheres (Standard)	1033.434 665	cm of water (0 °C)
atmospheres (Standard)	1033.289 446	cm of water (4 °C)
atmospheres (Standard)	1.013 25 × 10^6	dynes/sq centimetre
atmospheres (Standard)	33.900 572 39	feet of water (4 °C)
atmospheres (Standard)	1033.227 453	grams(force)/sq cm
atmospheres (Standard)	29.921 255 58	in of mercury (0 °C)
atmospheres (Standard)	1.033 227 453	kgf/square centimetre
atmospheres (Standard)	10 332.274 53	kgf/square metre
atmospheres (Standard)	* 101.3250	kilonewtons/sq metre
atmospheres (Standard)	1013.250	millibars
atmospheres (Standard)	759.999 891 8	mm of mercury (0 °C)
atmospheres (Standard)	101 325.0	newtons/square metre
atmospheres (Standard)	235.135 178 1	ounces/square inch
atmospheres (Standard)	101 325.0	pascals
atmospheres (Standard)	14.695 948 63	pounds(force)/sq inch
atmospheres (Standard)	0.006 560 691	tons(long)/square inch
atmospheres (Standard)	1.058 108 301	tons(short)/square foot
atmospheres (Standard)	* 760.0	torrs

Pressure or Stress

MULTIPLY	BY	TO OBTAIN
bars	0.986 923 267	atmospheres (Standard)
bars	1.0×10^6	baryes
bars	75.006 157 59	cm of mercury (0 °C)
bars	1.0×10^6	dynes/square centimetre
bars	33.457 263 64	feet of water (4 °C)
bars	33.499 417 27	feet of water (17 °C)
bars	1019.716 213	grams(force)/sq cm
bars	29.529 983 30	in of mercury (0 °C)
bars	1.019 716 213	kgf/square centimetre
bars	$1.019\ 72 \times 10^4$	kgf/square metre
bars	1000.0	millibars
bars	* 1.0×10^5	newtons/square metre
bars	2088.543 402	pounds(force)/sq foot
bars	14.503 773 63	pounds(force)/sq inch
baryes	9.8692×10^{-7}	atmospheres (Standard)
baryes	1.0×10^{-6}	bars
baryes	* 1.0	dynes/square centimetre
baryes	0.001 019 716	grams(force)/sq cm
baryes	0.0010	millibars
cm of mercury (0 °C)	0.013 157 897	atmospheres (Standard)
cm of mercury (0 °C)	0.013 332 239	bars
cm of mercury (0 °C)	13 332.238 74	dynes/sq centimetre
cm of mercury (0 °C)	0.446 060 227	feet of water (4 °C)
cm of mercury (0 °C)	0.446 622 229	feet of water (17 °C)
cm of mercury (0 °C)	0.393 700 787	in of mercury (0 °C)
cm of mercury (0 °C)	135.9510	kgf/square metre
cm of mercury (0 °C)	1333.223 874	newtons/square metre
cm of mercury (0 °C)	1333.223 874	pascals
cm of mercury (0 °C)	27.844 959 27	pounds(force)/sq foot
cm of mercury (0 °C)	0.193 367 773	pounds(force)/sq inch
cm of mercury (0 °C)	10.0	torrs
cm of water (0 °C)	0.000 967 647	atmospheres (Standard)
cm of water (0 °C)	980.468 369 0	dynes/square centimetre
cm of water (0 °C)	0.999 799 492	grams(force)/sq cm
cm of water (0 °C)	0.980 468 37	millibars
cm of water (0 °C)	98.046 836 90	newtons/square metre
cm of water (0 °C)	0.014 220 491	pounds(force)/sq inch
cm of water (4 °C)	0.000 967 783	atmospheres (Standard)
cm of water (4 °C)	980.606 163 6	dynes/square centimetre
cm of water (4 °C)	0.999 940 004	grams(force)/sq cm
cm of water (4 °C)	0.980 606 164	millibars

Pressure or Stress

MULTIPLY	BY	TO OBTAIN
cm of water (4 °C)	98.060 616 36	newtons/square metre
cm of water (4 °C)	0.014 222 490	pounds(force)/sq inch
dynes/sq centimetre	9.8692×10^{-7}	atmospheres (Standard)
dynes/sq centimetre	1.0×10^{-6}	bars
dynes/sq centimetre	1.0	baryes
dynes/sq centimetre	7.5006×10^{-5}	cm of mercury (0 °C)
dynes/sq centimetre	0.001 019 777	cm of water (4 °C)
dynes/sq centimetre	0.001 019 716	grams(force)/sq cm
dynes/sq centimetre	2.9530×10^{-5}	in of mercury (0 °C)
dynes/sq centimetre	0.000 401 487	in of water (4 °C)
dynes/sq centimetre	0.010 197 162	kgf/square metre
dynes/sq centimetre	0.0010	millibars
dynes/sq centimetre	0.10	newtons/square metre
dynes/sq centimetre	0.000 466 645	poundals/square inch
dynes/sq centimetre	0.000 014 504	pounds(force)/sq inch
feet of water (17 °C)	0.029 460 908	atmospheres (Standard)
feet of water (17 °C)	0.029 851 266	bars
feet of water (17 °C)	2.239 028 726	cm of mercury (0 °C)
feet of water (17 °C)	0.881 507 372	in of mercury (0 °C)
feet of water (17 °C)	0.030 439 819	kgf/sq centimetre
feet of water (17 °C)	304.398 194 4	kgf/square metre
feet of water (17 °C)	2.985 126 553	kilonewtons/sq metre
feet of water (17 °C)	2985.126 553	newtons/square metre
feet of water (17 °C)	62.345 663 67	pounds(force)/sq foot
feet of water (17 °C)	0.432 955 998	pounds(force)/sq inch
feet of water (17 °C)	0.000 193 284	tons(long)/square inch
feet of water (17 °C)	0.031 172 832	tons(short)/square foot
feet of water (4 °C)	0.029 498 027	atmospheres (Standard)
feet of water (4 °C)	0.029 888 876	bars
feet of water (4 °C)	2.241 849 734	cm of mercury (0 °C)
feet of water (4 °C)	29 888.875 87	dynes/square centimetre
feet of water (4 °C)	30.478 171 31	grams(force)/sq cm
feet of water (4 °C)	0.882 618 006	in of mercury (0 °C)
feet of water (4 °C)	0.030 478 171	kgf/sq centimetre
feet of water (4 °C)	304.781 713 1	kgf/square metre
feet of water (4 °C)	2.988 887 587	kilonewtons/sq metre
feet of water (4 °C)	2988.887 587	newtons/square metre
feet of water (4 °C)	62.424 214 49	pounds(force)/sq foot
feet of water (4 °C)	0.433 501 490	pounds(force)/sq inch
feet of water (4 °C)	0.000 193 527	tons(long)/square inch
feet of water (4 °C)	0.031 212 107	tons(short)/square foot

Pressure or Stress

MULTIPLY	BY	TO OBTAIN
grams(force)/square cm	0.000 967 841	atmospheres (Standard)
grams(force)/square cm	0.000 980 665	bars
grams(force)/square cm	0.073 555 914	cm of mercury (0 °C)
grams(force)/square cm	980.6650	dynes/square centimetre
grams(force)/square cm	0.028 959 021	in of mercury (0 °C)
grams(force)/square cm	10.0	kgf/square metre
grams(force)/square cm	0.735 559 135	mm of mercury (0 °C)
grams(force)/square cm	0.457 622 534	poundals/square inch
grams(force)/square cm	2.048 161 415	pounds(force)/sq foot
grams(force)/square cm	0.014 223 343	pounds(force)/sq inch
inches of water (17 °C)	0.002 455 076	atmospheres (Standard)
inches of water (17 °C)	0.073 458 948	in of mercury (0 °C)
inches of water (17 °C)	25.366 516 20	kgf/square metre
inches of water (17 °C)	2.487 605 461	millibars
inches of water (17 °C)	1.865 857 272	mm of mercury (0 °C)
inches of water (17 °C)	248.760 546 1	newtons/square metre
inches of water (17 °C)	0.577 274 664	ounces/square inch
inches of water (17 °C)	5.195 471 975	pounds(force)/sq foot
inches of water (17 °C)	0.036 079 666	pounds(force)/sq inch
inches of water (4 °C)	0.002 458 169	atmospheres (Standard)
inches of water (4 °C)	2490.739 656	dynes/square centimetre
inches of water (4 °C)	0.073 551 50	in of mercury (0 °C)
inches of water (4 °C)	0.002 539 848	kgf/square centimetre
inches of water (4 °C)	25.398 476 10	kgf/square metre
inches of water (4 °C)	2.490 739 656	millibars
inches of water (4 °C)	1.868 208 112	mm of mercury (0 °C)
inches of water (4 °C)	249.073 065 6	newtons/square metre
inches of water (4 °C)	83.232 286 01	ounces/square foot
inches of water (4 °C)	0.578 001 986	ounces/square inch
inches of water (4 °C)	5.202 017 875	pounds(force)/sq foot
inches of water (4 °C)	0.036 125 124	pounds(force)/sq inch
in of mercury (0 °C)	0.033 421 057	atmospheres (Standard)
in of mercury (0 °C)	0.033 863 886	bars
in of mercury (0 °C)	33 863.886 40	dynes/square centimetre
in of mercury (0 °C)	1.132 992 975	feet of water (4 °C)
in of mercury (0 °C)	1.134 420 461	feet of water (17 °C)
in of mercury (0 °C)	926.268 396 7	ft of air (1 atm, 17 °C)
in of mercury (0 °C)	34.531 554 0	grams(force)/sq cm
in of mercury (0 °C)	13.597 826 47	in of water (0 °C)
in of mercury (0 °C)	13.595 915 70	in of water (4 °C)
in of mercury (0 °C)	345.315 540	kgf/square metre

Pressure or Stress

MULTIPLY	BY	TO OBTAIN
in of mercury (0 °C)	0.034 531 554	kg/sq centimetre
in of mercury (0 °C)	3.386 388 640	kilonewtons/sq metre
in of mercury (0 °C)	25.40	mm of mercury (0 °C)
in of mercury (0 °C)	3386.388 640	newtons/square metre
in of mercury (0 °C)	7.858 466 283	ounces/square inch
in of mercury (0 °C)	70.726 196 55	pounds(force)/sq foot
in of mercury (0 °C)	0.491 154 143	pounds(force)/sq inch
in of mercury (60 °F)	0.033 326 804	atmospheres (Standard)
in of mercury (60 °F)	33 768.384 08	dynes/square centimetre
in of mercury (60 °F)	34.434 168 873	grams(force)/sq cm
in of mercury (60 °F)	25.328 367 37	mm of mercury (0 °C)
in of mercury (60 °F)	7.836 303 972	ounces/square inch
in of mercury (60 °F)	70.526 735 76	pounds(force)/sq foot
kgf/square centimetre	1.0	at (metric atmospheres)
kgf/square centimetre	0.967 841 105	atmospheres (Standard)
kgf/square centimetre	0.980 665 0	bars
kgf/square centimetre	73.555 913 53	cm of mercury (0 °C)
kgf/square centimetre	980 665.0	dynes/sq centimetre
kgf/square centimetre	32.810 367 45	feet of water (4 °C)
kgf/square centimetre	28.959 021 07	in of mercury (0 °C)
kgf/square centimetre	98.066 50	kilonewtons/sq metre
kgf/square centimetre	1.0	kilopond/sq centimetre
kgf/square centimetre	98 066.50	pascals
kgf/square centimetre	2048.161 415	pounds(force)/sq foot
kgf/square centimetre	14.223 343 16	pounds(force)/sq inch
kgf/square metre	9.6784×10^{-5}	atmospheres (Standard)
kgf/square metre	9.8067×10^{-5}	bars
kgf/square metre	0.007 355 591	cm of mercury (0 °C)
kgf/square metre	98.066 50	dynes/square centimetre
kgf/square metre	0.003 281 037	feet of water (4 °C)
kgf/square metre	0.003 285 171	feet of water (17 °C)
kgf/square metre	0.10	grams(force)/sq cm
kgf/square metre	0.002 895 902	in of mercury (0 °C)
kgf/square metre	0.098 066 50	millibars
kgf/square metre	0.073 555 914	mm of mercury (0 °C)
kgf/square metre	9.806 650	newtons/square metre
kgf/square metre	9.806 650	pascals
kgf/square metre	0.204 816 142	pounds(force)/sq foot
kgf/square metre	0.001 422 334	pounds(force)/sq inch
kgf/square millimetre	1.0×10^{6}	kgf/square metre
kp/sq centimetre	1.0	kgf/square centimetre

Pressure or Stress

MULTIPLY	BY	TO OBTAIN
kips/square inch	1000.0	pounds(force)/sq inch
metres of Hg (0.0 °C)	1.315 789 661	atmospheres (Standard)
metres of Hg (0.0 °C)	44.662 222 88	feet of water (17 °C)
metres of Hg (0.0 °C)	39.370 078 74	in of mercury (0 °C)
metres of Hg (0.0 °C)	1.359 510	kg/square centimetre
metres of Hg (0.0 °C)	19.336 777 27	pounds(force)/sq inch
metres of water (0 °C)	0.098 060 616	bars
metres of water (0 °C)	9806.061 636	newtons/metre squared
milli-atmospheres	1.013 250	millibars
milli-atmosperes (Std)	101.3250	newtons/metre squared
millibars	0.000 986 923	atmospheres (Standard)
millibars	0.0010	bars
millibars	1000.0	baryes
millibars	1000.0	dynes/square centimetre
millibars	1.019 716 213	grams(force)/sq cm
millibars	0.029 529 983	in of mercury (0 °C)
millibars	100.0	newtons/square metre
millibars	2.088 543 402	pounds(force)/sq foot
millibars	0.014 503 774	pounds(force)/sq inch
mm of mercury (0 °C)	0.001 315 790	atmospheres (Standard)
mm of mercury (0 °C)	0.001 333 224	bars
mm of mercury (0 °C)	1333.223 874	dynes/sq centimetre
mm of mercury (0 °C)	1.359 510	grams(force)/sq cm
mm of mercury (0 °C)	13.595 10	kgf/square metre
mm of mercury (0 °C)	1.333 223 874	millibars
mm of mercury (0 °C)	133.322 387 4	newtons/square metre
mm of mercury (0 °C)	2.784 495 926	pounds(force)/sq foot
mm of mercury (0 °C)	0.019 336 777	pounds(force)/sq inch
mm of mercury (0 °C)	1.000 000 142	torrs
mm of water (0 °C)	0.098 046 837	millibars
mm of water (0 °C)	9.804 683 690	newtons/square metre
newtons/sq millimetre	10.0	bars
newtons/sq millimetre	1.0×10^6	newtons/square metre
newtons/sq millimetre	145.037 736 3	pounds(force)/sq inch
newtons/square metre	* 10.0	dynes/square centimetre
newtons/square metre	0.010	millibars
newtons/square metre	1.0	pascals
newtons/square metre	0.000 145 038	pounds(force)/sq inch
newtons/square metre	6.4749×10^{-8}	tons(long)/square inch
ounces(force)/sq inch	0.004 252 873	atmospheres (Standard)
ounces(force)/sq inch	4.399 985 241	cm of water (17 °C)

Pressure or Stress

MULTIPLY	BY	TO OBTAIN
ounces(force)/sq inch	4309.223 352	dynes/sq centimetre
ounces(force)/sq inch	4.394 184 917	grams(force)/sq cm
ounces(force)/sq inch	1.732 277 653	inches water (17 °C)
ounces(force)/sq inch	1.730 097 861	inches water (4 °C)
ounces(force)/sq inch	0.127 251 294	in of mercury (0 °C)
ounces(force)/sq inch	9.0	pounds(force)/sq foot
ounces(force)/sq inch	0.062 50	pounds(force)/sq inch
pascals	* 1.0	newtons/square metre
poundals/square foot	0.014 881 640	millibars
poundals/square foot	1.488 163 958	newtons/metre squared
poundals/square inch	2.142 956 10	millibars
poundals/square inch	214.295 610	newtons/square metre
pounds(force)/sq foot	0.000 472 541	atmospheres (Standard)
pounds(force)/sq foot	0.000 478 803	bars
pounds(force)/sq foot	0.035 913 143	cm of mercury (0 °C)
pounds(force)/sq foot	478.802 594 7	dynes/sq centimetre
pounds(force)/sq foot	0.016 039 608	feet of water (17 °C)
pounds(force)/sq foot	13.138 956 18	ft of air (1 atm, 17 °C)
pounds(force)/sq foot	0.488 242 769	grams(force)/sq cm
pounds(force)/sq foot	0.014 139 033	in of mercury (0 °C)
pounds(force)/sq foot	0.192 233 096	in of water (4 °C)
pounds(force)/sq foot	4.882 427 686	kgf/square metre
pounds(force)/sq foot	0.478 802 595	millibars
pounds(force)/sq foot	0.359 131 429	mm of mercury (0 °C)
pounds(force)/sq foot	47.880 259 47	newtons/square metre
pounds(force)/sq foot	0.006 944 444	pounds(force)/sq inch
pounds(force)/sq inch	0.068 045 965	atmospheres (Standard)
pounds(force)/sq inch	0.068 947 574	bars
pounds(force)/sq inch	5.171 492 575	cm of mercury (0 °C)
pounds(force)/sq inch	70.311 177 11	cm of water (4 °C)
pounds(force)/sq inch	68 947.573 61	dynes/sq centimetre
pounds(force)/sq inch	2.309 703 538	feet of water (17 °C)
pounds(force)/sq inch	70.306 958 66	grams(force)/sq cm
pounds(force)/sq inch	2.036 020 697	in of mercury (0 °C)
pounds(force)/sq inch	27.681 565 78	in of water (4 °C)
pounds(force)/sq inch	0.070 306 959	kgf/sq centimetre
pounds(force)/sq inch	6.894 757 361	kilonewtons/sq metre
pounds(force)/sq inch	51.714 925 74	mm of mercury (0 °C)
pounds(force)/sq inch	6894.757 361	newtons/square metre
pounds(force)/sq inch	144.0	pounds(force)/sq foot
tons(long)/square foot	1.058 492 783	atmospheres (Standard)

Pressure or Stress

MULTIPLY	BY	TO OBTAIN
tons(long)/square foot	$1.075\ 52 \times 10^6$	dynes/sq centimetre
tons(long)/square foot	1093.663 802	grams(force)/sq cm
tons(long)/square foot	2240.0	pounds(force)/sq foot
tons(long)/square foot	15.555 555 56	pounds(force)/sq inch
tons(long)/square inch	152.422 960 7	atmospheres (Standard)
tons(long)/square inch	154.442 564 9	bars
tons(long)/square inch	$1.544\ 43 \times 10^8$	dynes/sq centimetre
tons(long)/square inch	157 487.5874	grams(force)/sq cm
tons(long)/square inch	$1.544\ 43 \times 10^7$	newtons/square metre
tons(long)/square inch	322 560.0	pounds(force)/sq foot
tons(long)/square inch	2240.0	pounds(force)/sq inch
tons(short)/square foot	0.945 082 842	atmospheres (Standard)
tons(short)/square foot	957 605.1894	dynes/sq centimetre
tons(short)/square foot	976.485 537 3	grams(force)/sq cm
tons(short)/square foot	9764.855 373	kgf/square metre
tons(short)/square foot	95 760.518 94	newtons/square metre
tons(short)/square foot	2000.0	pounds(force)/sq foot
tons(short)/square foot	13.888 888 89	pounds(force)/sq inch
tons(short)/square inch	136.091 929 1	atmospheres
tons(short)/square inch	$1.378\ 95 \times 10^8$	dynes/sq centimetre
tons(short)/square inch	140 613.9173	grams(force)/sq cm
tons(short)/square inch	1.406 139 173	kgf/square millimetre
tons(short)/square inch	2000.0	pounds(force)/sq inch
tons(short)/square inch	288 000.0	pounds/square foot
torrs	0.999 999 858	mm of mercury (0 °C)
torrs	133.322 368 4	pascals

SOUND
Dimensions: None

MULTIPLY	BY	TO OBTAIN
bels	10.0	decibels
decibels	0.10	bels
nepers	8.6860	decibels

SPECIFIC ENERGY (Specific heat)
Dimensions: L^2/T^2

MULTIPLY	BY	TO OBTAIN
Btu/pound °F	4186.80	joules/kilogram K
Btu/pound °F	4.186 80	kilojoules/kg °C

Specific Energy (Specific heat)

MULTIPLY	BY	TO OBTAIN
Btu/pound °R	4186.80	joules/kilogram K
Btu/pound °R	4.186 80	kilojoules/kg °C
calories/gram °C	4186.80	joules/kilogram K
calories/gram °C	4.186 80	kilojoules/kg K
Chu/pound °C	4186.80	joules/kilogram K
Chu/pound °C	4.186 80	kilojoules/kg °C
joules/kilogram °C	1.0000	joules/kilogram K
joules/kilogram °C	0.0010	kilojoules/kg °C
kilocalories/kg °C	4186.80	joules/kilogram K
kilocalories/kg °C	4.186 80	kilojoules/kg °C
kilojoules/kg °C	1000.0	joules/kilogram K

SPECIFIC VOLUME
Dimensions: L^3/M

MULTIPLY	BY	TO OBTAIN
cubic centimetres/gram	0.016 018 464	cubic foot/pound
cubic centimetres/gram	0.0010	metres cubed/kilogram
cubic feet/kilogram	0.028 316 847	cubic metres/kilogram
cubic feet/pound	62.427 959 95	cubic centimetres/gram
cubic feet/pound	0.062 427 960	cubic metres/kilogram
cubic metres/kilogram	99.776 373 65	gallons (UK)/pound
litres/gram	1.0000	cubic metres/kilogram
litres/kilogram	0.0010	cubic metres/kilogram

SURFACE TENSION
Dimensions: M/T^2

MULTIPLY	BY	TO OBTAIN
dynes/centimetre	1.0	ergs/square centimetre
dynes/centimetre	0.010	ergs/square millimetre
dynes/centimetre	0.001 019 716	grams(f)/centimetre
dynes/centimetre	1.0	millinewtons/metre
dynes/centimetre	0.0010	newtons/metre
dynes/centimetre	0.000 183 719	poundals/inch
millinewtons/metre	0.0010	newtons/metre
poundals/inch	5.443 108 492	newtons/metre
pounds(force)/inch	175.126 837 0	newtons/metre

TEMPERATURE
Dimensions: None

MULTIPLY	BY	TO OBTAIN
Celsius temperature (C)	9/5 (then add 32)	Fahrenheit temp. (F)
Rankin temperature (R)	5/9	Kelvin temperature (K)
temperature (°C + 17.78)	1.80	temperature (°F)
temperature (°F − 32)	5/9	temperature (°C)
temperature (K)	1.80	temperature (°R)
temperature (°R)	5/9	temperature (K)
temperature (°C + 273.15)	1.0	temperature (K)
temperature (°C + 273.15)	1.80	temperature (°R)
temperature (°F + 459.67)	5/9	temperature (K)
temperature (°F + 459.67)	1.0	temperature (°R)
temperature (K − 255.37)	1.80	temperature (°F)
temperature (K − 273.15)	1.0	temperature (°C)
temperature (°R − 459.67)	1.0	temperature (°F)
temperature (°F − 491.67)	5/9	temperature (°C)

TEMPERATURE DIFFERENCE
Dimensions: None

MULTIPLY	BY	TO OBTAIN
degrees Celsius (C)	1.80	degrees Fahrenheit (F)
degrees Celsius (C)	1.0	degrees Kelvin (K)
degrees Celsius (C)	1.80	degrees Rankin (R)
degrees Fahrenheit (F)	5/9	degrees Celsius (C)
degrees Fahrenheit (F)	5/9	degrees Kelvin (K)
degrees Fahrenheit (F)	1.0	degrees Rankin (R)
degrees Kelvin (K)	1.0	degrees Celsius (C)
degrees Kelvin (K)	1.80	degrees Fahrenheit (F)
degrees Kelvin (K)	1.80	degrees Rankin (R)
degrees Rankin (R)	5/9	degrees Celsius (C)
degrees Rankin (R)	1.0	degrees Fahrenheit (F)
degrees Rankin (R)	5/9	degrees Kelvin (K)

THERMAL CONDUCTIVITY
Dimensions: ML/T^3

MULTIPLY	BY	TO OBTAIN
Btu/hour foot °F	1.730 734 684	joules/second metre K
Btu/hour foot °F	1.730 734 684	watts/metre °C

Thermal Conductivity

MULTIPLY	BY	TO OBTAIN
Btu/hour foot °F	1.730 734 684	watts/metre K
Btu/hour ft² (F/inch)	0.144 227 890	joules/second metre K
Btu/hour ft² (F/inch)	0.144 227 890	watts/metre °C
calories/second cm °C	418.680	joules/second metre K
calories/second cm °C	418.680	watts/metre °C
Chu/hour foot °C	1.730 734 684	joules/second metre K
Chu/hour foot °C	1.730 734 684	watts/metre °C
kilocal/hour foot °C	3.815 616 798	joules/second metre K
kilocal/hour foot °C	3.815 616 798	watts/metre °C
kilocal/hour metre °C	1.1630	joules/second metre K
kilocal/hour metre °C	1.1630	watts/metre °C
kilocal/hour metre °C	1.1630	watts/metre K

TIME
Dimensions: T

MULTIPLY	BY	TO OBTAIN
days (mean solar)	1.002 737 803	days (sidereal)
days (mean solar)	* 24.0	hours (mean solar)
days (mean solar)	24.065 707 27	hours (sidereal)
days (mean solar)	1440.0	minutes (mean solar)
days (mean solar)	86 400.0	seconds (mean solar)
days (mean solar)	0.002 739 726	years (calendar)
days (mean solar)	0.002 737 803	years (sidereal)
days (mean solar)	0.002 737 909	years (tropical)
days (sidereal)	0.997 269 672	days (mean solar)
days (sidereal)	23.934 472 12	hours (mean solar)
days (sidereal)	* 24.0	hours (sidereal)
days (sidereal)	1436.068 327	minutes (mean solar)
days (sidereal)	1440.0	minutes (sidereal)
days (sidereal)	86 164.099 62	seconds (mean solar)
days (sidereal)	86 400.0	seconds (sidereal)
days (sidereal)	0.002 732 246	years (calendar)
days (sidereal)	0.002 730 328	years (sidereal)
days (sidereal)	0.002 730 434	years (tropical)
hours (mean solar)	1/24	days (mean solar)
hours (mean solar)	0.041 780 742	days (sidereal)
hours (mean solar)	1.002 737 803	hours (sidereal)
hours (mean solar)	* 60.0	minutes (mean solar)
hours (mean solar)	60.164 268 18	minutes (sidereal)
hours (mean solar)	3600.0	seconds (mean solar)

Time

MULTIPLY	BY	TO OBTAIN
hours (mean solar)	3609.856 091	seconds (sidereal)
hours (mean solar)	1/168	weeks (mean calendar)
hours (sidereal)	0.415 529 03	days (mean solar)
hours (sidereal)	1/24	days (sidereal)
hours (sidereal)	0.997 269 672	hours (mean solar)
hours (sidereal)	59.836 180 33	minutes (mean solar)
hours (sidereal)	* 60.0	minutes (sidereal)
minutes (mean solar)	1/1440	days (mean solar)
minutes (mean solar)	0.000 696 346	days (sidereal)
minutes (mean solar)	1/60	hours (mean solar)
minutes (mean solar)	0.016 712 297	hours (sidereal)
minutes (mean solar)	1.002 737 803	minutes (sidereal)
minutes (mean solar)	* 60.0	seconds
minutes (mean solar)	1/100 80	weeks
minutes (sidereal)	0.000 692 548	days (mean solar)
minutes (sidereal)	0.997 269 57	minutes (mean solar)
minutes (sidereal)	2.2769×10^{-5}	months (mean calendar)
minutes (sidereal)	* 60.0	seconds (sidereal)
months (lunar)	** 29.530 589 04	days (mean solar)
months (lunar)	708.734 137 0	hours (mean solar)
months (lunar)	42 524.048 22	minutes (mean solar)
months (lunar)	$2.551 44 \times 10^{6}$	seconds (mean solar)
months (lunar)	4.218 655 576	weeks (mean calendar)
months (mean calendar)	30.416 666 67	days (mean solar)
months (mean calendar)	* 730.0	hours (mean solar)
months (mean calendar)	1.030 005 417	months (lunar)
months (mean calendar)	4.345 238 096	weeks (mean calendar)
months (mean calendar)	1/12	years (calendar)
months (mean calendar)	0.083 274 846	years (sidereal)
months (mean calendar)	0.083 278 073	years (tropical)
seconds (mean solar) †	1/864 00	days (mean solar)
seconds (mean solar)	1/861 64	days (sidereal)
seconds (mean solar)	1/3600	hours (mean solar)
seconds (mean solar)	0.000 278 538	hours (sidereal)
seconds (mean solar)	1/60	minutes (mean solar)
seconds (mean solar)	0.016 712 297	minutes (sidereal)
seconds (mean solar)	1.002 737 803	seconds (sidereal)
seconds (mean solar)	1/604 800	weeks (mean solar)
seconds (sidereal)	1.1543×10^{-5}	days (mean solar)

† SI unit

Time

MULTIPLY	BY	TO OBTAIN
seconds (sidereal)	1.1574×10^{-5}	days (sidereal)
seconds (sidereal)	0.000 277 019	hours (mean solar)
seconds (sidereal)	1/3600	hours (sidereal)
seconds (sidereal)	0.016 621 161	minutes (mean solar)
seconds (sidereal)	1/60	minutes (sidereal)
seconds (sidereal)	0.997 269 672	seconds (mean solar)
years (calendar)	* 365.0	days (mean solar)
years (calendar)	8760.0	hours (mean solar)
years (calendar)	525 600.0	minutes (mean solar)
years (calendar)	12.360 065 0	months (lunar)
years (calendar)	* 12.0	months (mean calendar)
years (calendar)	$3.153\ 60 \times 10^7$	seconds (mean solar)
years (calendar)	52.142 857 14	weeks (mean calendar)
years (calendar)	0.999 298 153	years (sidereal)
years (calendar)	0.999 336 881	years (tropical)
years (calendar, leap)	366.0	days (mean solar)
years (sidereal)	** 365.256 354 2	days (mean solar)
years (sidereal)	366.256 354 2	days (sidereal)
years (sidereal)	8790.152 498	hours (sidereal)
years (sidereal)	527 409.149 9	minutes (sidereal)
years (sidereal)	3.1645×10^7	seconds (sidereal)
years (sidereal)	1.000 702 340	years (calendar)
years (sidereal)	1.000 038 755	years (tropical)
years (tropical)	** 365.242 199 1	days (mean solar)
years (tropical)	366.242 199 1	days (sidereal)
years (tropical)	8765.812 776	hours (mean solar)
years (tropical)	8789.812 776	hours (sidereal)
years (tropical)	525 948.7666	minutes (mean solar)
years (tropical)	12.007 962 71	months (mean calendar)
years (tropical)	$3.155\ 69 \times 10^7$	seconds (mean solar)
years (tropical)	$3.164\ 33 \times 10^7$	seconds (sidereal)
years (tropical)	52.177 457 0	weeks (mean calendar)
years (tropical)	1.000 663 559	years (calendar)
years (tropical)	0.999 961 246	years (sidereal)

TORQUE
Dimensions: ML^2/T^2

MULTIPLY	BY	TO OBTAIN
foot pounds	32.174 048 56	foot poundals
foot pounds	13 825.495 43	gram(force) centimetres

Torque

MULTIPLY	BY	TO OBTAIN
foot pounds	192.0	inch ounces
foot pounds	12.0	inch pounds
foot pounds	13.825 495 43	kgf centimetres
foot pounds	0.138 254 954	kilogram(force) metres
foot pounds	135.581 794 8	newton centimetres
foot pounds	1.355 817 948	newton metres (joules)
gram(force) centimetres	723 301.3853	foot pounds
gram(force) centimetres	0.013 887 387	inch ounces
gram(force) centimetres	8.6796×10^{-4}	inch pounds
gram(force) centimetres	0.0010	kgf centimetres
gram(force) centimetres	1.0×10^{-5}	kilogram(force) metres
gram(force) centimetres	0.009 806 65	newton centimetres
gram(force) centimetres	9.8067×10^{-5}	newton metres
inch ounces	0.005 208 333	foot pounds
inch ounces	72.007 788 71	gram(force) centimetres
inch ounces	0.062 50	inch pounds
inch ounces	0.072 007 789	kgf centimetres
inch ounces	0.000 720 078	kilogram(force) metres
inch ounces	0.706 155 181	newton centimetres
inch ounces	0.007 061 552	newton metres
inch pounds	1/12	foot pounds
inch pounds	1152.124 620	gram(force) centimetres
inch pounds	16.0	inch ounces
inch pounds	1.152 124 620	kgf centimetres
inch pounds	0.011 521 246	kilogram(force) metres
inch pounds	11.298 482 90	newton centimetres
inch pounds	0.112 984 829	newton metres
kgf centimetres	0.072 330 139	foot pounds
kgf centimetres	1000.0	gram(force) centimetres
kgf centimetres	13.887 386 60	inch ounces
kgf centimetres	0.867 961 662	inch pounds
kgf centimetres	0.0100	kilogram(force) metres
kgf centimetres	9.806 650	newton centimetres
kgf centimetres	0.098 066 50	newton metres
kilogram(force) metres	7.233 013 855	foot pounds
kilogram(force) metres	1.0×10^{5}	gram(force) centimetres
kilogram(force) metres	1388.738 646	inch ounces
kilogram(force) metres	86.796 216 624	inch pounds
kilogram(force) metres	100.0	kgf centimetres
kilogram(force) metres	980.6650	newton centimetres
kilogram(force) metres	9.806 650	newton metres

Torque

MULTIPLY	BY	TO OBTAIN
newton centimetres	0.007 375 621	foot pounds
newton centimetres	101.971 621 3	gram(force) centimetres
newton centimetres	1.416 119 313	inch ounces
newton centimetres	0.088 507 457	inch pounds
newton centimetres	0.101 971 621	kgf centimetres
newton centimetres	0.001 019 716	kilogram(force) metres
newton centimetres	0.010	newton metres
newton metres	1.0×10^7	dyne centimetres
newton metres	0.737 562 150	foot pounds
newton metres	10 197.162 13	gram(force) centimetres
newton metres	141.611 932 7	inch ounces
newton metres	8.850 745 705	inch pounds
newton metres	10.197 162 13	kgf centimetres
newton metres	0.101 971 621	kilogram(force) metres
newton metres	100.0	newton centimetres

VELOCITY (Angular)
Dimensions: 1/T

MULTIPLY	BY	TO OBTAIN
degrees (angle)/minute	1/60	degrees (angle)/second
degrees (angle)/minute	0.000 290 888	radians/second
degrees (angle)/minute	1/216 00	revolutions/second
degrees (angle)/second	0.017 453 293	radians/second
degrees (angle)/second	1/6	revolutions/minute
degrees (angle)/second	1/360	revolutions/second
radians/second	57.295 779 51	degrees/second
radians/second	9.549 296 586	revolutions/minute
radians/second	0.159 154 943	revolutions/second
revolutions/min (RPM)	6.0	degrees/second
revolutions/min (RPM)	0.104 719 755	radians/second
revolutions/min (RPM)	0.016 666 667	revolutions/second
revolutions/second	360.0	degrees/second
revolutions/second	6.283 185 308	radians/second
revolutions/second	60.0	revolutions/minute

VELOCITY (Linear)
Dimensions: L/T

MULTIPLY	BY	TO OBTAIN
centimetres/second	1.968 503 937	feet/minute
centimetres/second	0.032 808 399	feet/second

Velocity (Linear)

MULTIPLY	BY	TO OBTAIN
centimetres/second	0.0360	kilometres/hour
centimetres/second	0.000 60	kilometres/minute
centimetres/second	0.019 438 445	knots (International)
centimetres/second	0.60	metres/minute
centimetres/second	0.010	metres/second
centimetres/second	0.022 369 363	miles/hour
centimetres/second	0.000 372 823	miles/minute
diam (inches) × RPM	0.261 799 388	feet/min (belt speed)
diam (inches) × RPM	* 4.787 787 204	metres/h (belt speed)
feet/hour	30.480	centimetres/hour
feet/hour	0.5080	centimetres/minute
feet/hour	0.008 466 667	centimetres/second
feet/hour	1/60	feet/minute
feet/hour	0.20	inches/minute
feet/hour	0.000 304 80	kilometres/hour
feet/hour	5.080×10^{-6}	kilometres/minute
feet/hour	0.000 164 579	knots (International)
feet/hour	8.4667×10^{-5}	metres/second
feet/hour	0.000 189 394	miles/hour
feet/hour	3.1566×10^{-6}	miles/minute
feet/hour	5.2609×10^{-8}	miles/second
feet/hour	0.084 666 667	millimetres/second
feet/minute	0.5080	centimetres/second
feet/minute	1/60	feet/second
feet/minute	0.018 288 0	kilometres/hour
feet/minute	0.009 874 730	knots (International)
feet/minute	0.304 80	metres/minute
feet/minute	0.005 080	metres/second
feet/minute	0.011 363 636	miles/hour
feet/second	30.480	centimetres/second
feet/second	60.0	feet/minute
feet/second	1.097 280	kilometres/hour
feet/second	0.018 288 0	kilometres/minute
feet/second	0.592 483 801	knots (International)
feet/second	18.2880	metres/minute
feet/second	* 0.304 80	metres/second
feet/second	0.681 818 182	miles/hour
feet/second	0.011 363 636	miles/minute
kilometres/hour	27.777 777 78	centimetres/second
kilometres/hour	3280.839 895	feet/hour
kilometres/hour	54.680 664 92	feet/minute

83

Velocity (Linear)

MULTIPLY	BY	TO OBTAIN
kilometres/hour	0.911 344 415	feet/second
kilometres/hour	0.539 956 804	knots (International)
kilometres/hour	16.666 666 67	metres/minute
kilometres/hour	0.277 777 778	metres/second
kilometres/hour	0.621 371 192	miles/hour
kilometres/minute	1666.666 667	centimetres/second
kilometres/minute	3280.839 895	feet/minute
kilometres/minute	60.0	kilometres/hour
kilometres/minute	32.397 408 21	knots (International)
kilometres/minute	37.282 271 53	miles/hour
kilometres/minute	0.621 371 192	miles/minute
knots (International)	51.444 444 44	centimetres/second
knots (International)	6076.115 486	feet/hour
knots (International)	101.268 591 4	feet/minute
knots (International)	1.687 809 857	feet/second
knots (International)	* 1.8520	kilometres/hour
knots (International)	30.866 666 67	metres/minute
knots (International)	0.514 444 444	metres/second
knots (International)	* 1.0	miles (nautical)/hour
knots (International)	1.150 779 448	miles (statute)/hour
knots (International)	2025.371 828	yards/hour
metres/hour	3.280 839 895	feet/hour
metres/hour	0.054 680 665	feet/minute
metres/hour	0.000 539 957	knots (International)
metres/hour	0.000 621 371	miles (statute)/hour
metres/minute	1.666 666 667	centimetres/second
metres/minute	3.280 839 895	feet/minute
metres/minute	0.054 680 665	feet/second
metres/minute	0.060	kilometres/hour
metres/minute	0.032 397 408	knots (International)
metres/minute	0.016 666 667	metres/second
metres/minute	0.037 282 272	miles (statute)/hour
metres/second	196.850 393 7	feet/minute
metres/second	3.280 839 895	feet/second
metres/second	3.60	kilometres/hour
metres/second	0.060	kilometres/minute
metres/second	60.0	metres/minute
metres/second	2.236 936 292	miles/hour
metres/second	0.037 282 272	miles/minute
miles/hour	44.7040	centimetres/second
miles/hour	5280.0	feet/hour

Velocity (Linear)

MULTIPLY	BY	TO OBTAIN
miles/hour	88.0	feet/minute
miles/hour	1.466 666 667	feet/second
miles/hour	1.609 344 0	kilometres/hour
miles/hour	0.026 822 40	kilometres/minute
miles/hour	0.868 976 242	knots (International)
miles/hour	26.822 40	metres/minute
miles/hour	0.447 040	metres/second
miles/hour	1/60	miles/minute
miles/minute	2682.240	centimetres/second
miles/minute	316 800.0	feet/hour
miles/minute	88.0	feet/second
miles/minute	1.609 344 0	kilometres/minute
miles/minute	52.138 574 51	knots (International)
miles/minute	1609.3440	metres/minute
miles/minute	60.0	miles/hour
miles/second	1609.3440	metres/second
millimetres/second	0.0010	metres/second

VISCOSITY (Absolute, dynamic)
Dimensions: M/LT

MULTIPLY	BY	TO OBTAIN
centipoise	* 0.010	grams/centimetre second
centipoise	* 1.0	mN seconds/sq metre
centipoise	0.010	poise
centipoise	2.419 088 287	pounds/foot hour
centipoise	0.000 671 969	pounds/foot second
grams/centimetre second	0.10	newton seconds/metre^2
grams/centimetre second	1.0	poise
grams/centimetre second	0.067 196 897	pounds/foot second
gf/metre second^2	1.0	mN seconds/sq metre
gf seconds/square metre	9.806 650	pascal seconds
gf/metre hour	0.277 777 778	mN seconds/sq metre
gf/metre hour	0.000 277 778	newton seconds/metre^2
gf/mm^2	9.806 650	Megapascals
gf/mm^2	9.806 650	newtons/sq millimetre
gf/foot hour	0.911 344 415	mN seconds/sq metre
gf/foot hour	0.000 911 344	newton seconds/metre^2
mN seconds/sq metre	1.0	centipoise
mN seconds/sq metre	0.0010	newton seconds/metre^2

Viscosity (Absolute, dynamic)

MULTIPLY	BY	TO OBTAIN
poise	100.0	centipoise
poise	* 1.0	dyne second/square cm
poise	* 1.0	grams/centimetre second
poise	0.10	newton seconds/metre^2
poise	241.908 828 7	pounds/foot hour
poise	0.067 196 897	pounds/foot second
pounds/foot hour	0.413 378 877	centipoise
pounds/foot hour	0.413 378 877	mN seconds/sq metre
pounds/foot hour	0.000 413 379	newton seconds/metre^2
pounds/foot hour	0.004 133 789	poise grams/second cm
pounds/foot hour	1/3600	pounds/foot second
pounds/foot second	1488.163 958	centipoise
pounds/foot second	1488.163 958	mN seconds/sq metre
pounds/foot second	1.488 163 958	newton seconds/metre^2
pounds/foot second	14.881 639 58	poise grams/second cm
pounds/foot second	3600.0	pounds/foot hour

VISCOSITY (Kinematic)
Dimensions: L^2/T

MULTIPLY	BY	TO OBTAIN
centistokes	1.0	sq millimetres/second
centistokes	0.010	stokes
litres/inch hour	10.936 132 98	sq millimetres/second
litres/inch hour	1.0936×10^{-5}	square metres/second
sq centimetres/second	100.0	sq millimetres/second
sq centimetres/second	0.00010	square metres/second
sq millimetres/second	1.0000×10^{-6}	square metres/second
square feet/hour	25.806 399 99	sq millimetres/second
square feet/hour	2.5806×10^{-5}	square metres/second
square feet/second	0.092 903 040	square metres/second
square metres/hour	277.777 777 78	sq millimetres/second
square metres/hour	0.000 277 778	square metres/second
stokes	1.0	CGS kinematic viscosity
stokes	1.0	cubic cm/gram (poise)
stokes	* 1.0	sq centimetres/second
stokes	0.155 000 310	square inch/second

VOLUME (Capacity)
Dimensions: L^3

MULTIPLY	BY	TO OBTAIN
acre feet	3.258 51 $\times 10^5$	barrels (US dry)
acre feet	43 560.0	cubic feet
acre feet	1233.481 838	cubic metres
acre feet	1613.333 333	cubic yards
acre feet	2.713 28 $\times 10^5$	gallons (UK)
acre feet	1.233 48 $\times 10^6$	litres
acre inches	27 154.285 34	barrels (US dry)
acre inches	3630.0	cubic feet
acre inches	102.790 153 1	cubic metres
acre inches	22 610.672 71	gallons (UK)
bags (UK)	* 3.0	bushels (UK)
barrels (lube oil, US)	0.158 987 297	cubic metres
barrels (lube oil, US)	34.972 315 77	gallons (UK)
barrels (lube oil, US)	* 42.0	gallons (US liquid)
barrels (UK)	* 1.5	bags (UK)
barrels (UK)	1.415 405 269	barrels (US dry)
barrels (UK)	1.372 514 20	barrels (US liquid)
barrels (UK)	* 4.5	bushels (UK)
barrels (UK)	4.644 255 348	bushels (US)
barrels (UK)	5.779 571 516	cubic feet
barrels (UK)	0.163 659 240	cubic metres
barrels (UK)	* 36.0	gallons (UK)
barrels (UK)	163.659 240	litres
barrels (UK wine)	* 31.50	gallons (UK)
barrels (US dry)	0.969 696 970	barrels (US liquid)
barrels (US dry)	3.281 219 485	bushels (US)
barrels (US dry)	4.083 333 333	cubic feet
barrels (US dry)	* 7056.0	cubic inches
barrels (US dry)	0.115 627 124	cubic metres
barrels (US dry)	26.249 755 88	gallons (US dry)
barrels (US dry)	104.999 023 5	quarts (US dry)
barrels (US oil)	5.614 583 333	cubic feet
barrels (US oil)	0.158 987 295	cubic metres
barrels (US oil)	* 42.0	gallons (US liquid)
barrels (US oil)	158.987 294 9	litres
barrels (US liquid)	0.832 674 185	barrels (UK wine)
barrels (US liquid)	1.031 250	barrels (US dry)
barrels (US liquid)	4.210 937 556	cubic feet

Volume (Capacity)

MULTIPLY	BY	TO OBTAIN
barrels (US liquid)	* 7276.50	cubic inches
barrels (US liquid)	0.119 240 473	cubic metres
barrels (US liquid)	26.229 236 82	gallons (UK)
barrels (US liquid)	* 31.5	gallons (US liquid)
barrels (US liquid)	119.240 471 2	litres
board feet (timber)	2359.737 216	cubic centimetres
board feet (timber)	0.083 333 333	cubic feet
board feet (timber)	* 144.0	cubic inches
buckets (UK)	18 184.360	cubic centimetres
buckets (UK)	* 4.0	gallons (UK)
buckets (UK)	18.184 360	litres
bushels (UK)	1/3	bags (UK)
bushels (UK)	1.032 056 744	bushels (US)
bushels (UK)	36 368.720	cubic centimetres
bushels (UK)	1.284 349 226	cubic feet
bushels (UK)	2219.355 463	cubic inch
bushels (UK)	0.036 368 720	cubic metres
bushels (UK)	3.636 872 0	dekalitres
bushels (UK)	* 8.0	gallons (UK)
bushels (UK)	0.363 687 20	hectolitres
bushels (UK)	36.368 720	litres
bushels (US)	0.304 764 739	barrels (US dry)
bushels (US)	0.968 938 972	bushels (UK)
bushels (US)	35 239.070 15	cubic centimetres
bushels (US)	1.244 456 018	cubic feet
bushels (US)	2150.419 999	cubic inches
bushels (US)	0.035 239 070	cubic metres
bushels (US)	0.046 090 964	cubic yards
bushels (US)	* 8.0	gallons (US dry)
bushels (US)	9.309 177 482	gallons (US liquid)
bushels (US)	35.239 070 15	litres
bushels (US)	1191.574 718	ounces (US fluid)
bushels (US)	* 4.0	pecks (US)
bushels (US)	64.0	pints (US dry)
bushels (US)	32.0	quarts (US dry)
bushels (US)	37.236 709 93	quarts (US liquid)
butts (UK)	13.545 744 76	bushels (US)
butts (UK)	16.857 083 59	cubic feet
butts (UK)	0.477 339 450	cubic metres
butts (UK)	* 105.0	gallons (UK)
butts (UK)	126.099 742 1	gallons (US liquid)

Volume (Capacity)

MULTIPLY	BY		TO OBTAIN
butts (UK)	*	2.0	hogsheads
centilitres		10.0	cubic centimetres
centilitres		0.610 237 441	cubic inches
centilitres		0.010	litres
centilitres		0.338 140 227	ounces (US fluid)
cord feet (timber)		0.1250	cords (timber)
cord feet (timber)	*	16.0	cubic feet
cord feet (timber)		0.453 069 545	cubic metres
cords (timber)		8.0	cord feet (timber)
cords (timber)	*	128.0	cubic feet
cords (timber)		3.624 556 364	cubic metres
cubic centimetres		0.000 423 776	board feet
cubic centimetres		2.7496×10^{-5}	bushels (UK)
cubic centimetres		2.8378×10^{-5}	bushels (US)
cubic centimetres		0.0010	cubic decimetres
cubic centimetres		3.5315×10^{-5}	cubic feet
cubic centimetres		0.061 023 744	cubic inches
cubic centimetres		1.0×10^{-6}	cubic metres
cubic centimetres		1.3080×10^{-6}	cubic yards
cubic centimetres		0.000 219 969	gallons (UK)
cubic centimetres		0.000 227 021	gallons (US dry)
cubic centimetres		0.000 264 172	gallons (US liquid)
cubic centimetres		0.007 039 016	gills (UK)
cubic centimetres		0.008 453 506	gills (US)
cubic centimetres	*	13.595 10	grams of Hg (0.0 °C)
cubic centimetres		13.585 237 28	grams of Hg (4.0 °C)
cubic centimetres		0.0010	litres
cubic centimetres		1.0	millilitres
cubic centimetres		0.035 195 080	ounces (UK fluid)
cubic centimetres		0.033 814 023	ounces (US fluid)
cubic centimetres		0.001 759 754	pints (UK)
cubic centimetres		0.001 816 166	pints (US dry)
cubic centimetres		0.002 113 376	pints (US liquid)
cubic centimetres		0.000 879 877	quarts (UK)
cubic centimetres		0.000 908 083	quarts (US dry)
cubic centimetres		0.001 056 688	quarts (US liquid)
cubic decametres		1.0×10^{6}	cubic decimetres
cubic decametres		35 314.666 72	cubic feet
cubic decametres		$6.102\ 37 \times 10^{7}$	cubic inches
cubic decametres		1000.0	cubic metres
cubic decametres		1.0×10^{6}	litres

Volume (Capacity)

MULTIPLY	BY	TO OBTAIN
cubic decimetres	1000.0	cubic centimetres
cubic decimetres	0.035 314 667	cubic feet
cubic decimetres	61.023 744 10	cubic inches
cubic decimetres	0.0010	cubic metres
cubic decimetres	0.001 307 951	cubic yards
cubic decimetres	1.0	litres
cubic feet	2.2957×10^{-5}	acre feet
cubic feet	12.0	board feet (timber)
cubic feet	0.778 604 432	bushels (UK)
cubic feet	0.803 563 955	bushels (US)
cubic feet	0.062 50	cord feet (timber)
cubic feet	0.007 812 500	cords (timber)
cubic feet	28 316.846 59	cubic centimetres
cubic feet	1728.0	cubic inches
cubic feet	0.028 316 847	cubic metres
cubic feet	1/27	cubic yards
cubic feet	6.228 835 460	gallons (UK)
cubic feet	6.428 511 644	gallons (US dry)
cubic feet	7.480 519 478	gallons (US liquid)
cubic feet	62.356 729 66	lb of water (15.18 °C)
cubic feet	62.345 663 68	lb of water (17.00 °C)
cubic feet	62.424 214 50	lb of water (4.00 °C)
cubic feet	28.316 865 9	litres
cubic feet	996.613 673 2	ounces (UK fluid)
cubic feet	957.506 493 8	ounces (US fluid)
cubic feet	49.830 683 68	pints (UK)
cubic feet	59.844 155 82	pints (US liquid)
cubic feet	25.714 046 58	quarts (US dry)
cubic feet	29.922 077 91	quarts (US liquid)
cubic inches	0.000 100 129	barrels (UK)
cubic inches	0.000 141 723	barrels (US dry)
cubic inches	1/144	board feet (timber)
cubic inches	0.000 450 581	bushels (UK)
cubic inches	0.000 465 025	bushels (US)
cubic inches	$1.061\ 03 \times 10^{5}$	circular mil feet
cubic inches	16.387 064 00	cubic centimetres
cubic inches	0.000 578 704	cubic feet
cubic inches	1.6387×10^{-5}	cubic metres
cubic inches	2.1434×10^{-5}	cubic yards
cubic inches	4.432 900 433	drams (US fluid)
cubic inches	0.003 604 650	gallons (UK)

Volume (Capacity)

MULTIPLY	BY	TO OBTAIN
ubic inches	0.003 720 203	gallons (US dry)
ubic inches	1/231	gallons (US liquid)
ubic inches	0.016 387 064	litres
ubic inches	16.387 064 00	millilitres
ubic inches	0.576 744 024	ounces (UK fluid)
ubic inches	0.554 112 554	ounces (US fluid)
ubic inches	0.001 860 102	pecks (US)
ubic inches	0.028 837 201	pints (UK)
ubic inches	0.029 761 628	pints (US dry)
ubic inches	0.034 632 035	pints (US liquid)
ubic inches	0.014 880 814	quarts (US dry)
ubic inches	0.017 316 017	quarts (US liquid)
ubic metres	0.000 810 713	acre feet
ubic metres	6.110 256 897	barrels (UK)
ubic metres	8.648 489 808	barrels (US dry)
ubic metres	8.386 414 360	barrels (US liquid)
ubic metres	27.496 156 04	bushels (UK)
ubic metres	28.377 593 27	bushels (US)
ubic metres	1.0×10^6	cubic centimetres
ubic metres	35.314 666 72	cubic feet
ubic metres	61 023.744 09	cubic inches
ubic metres	1.307 950 619	cubic yards
ubic metres	219.969 248 3	gallons (UK)
ubic metres	264.172 05	gallons (US liquid)
ubic metres	4.189 890 444	hogsheads
ubic metres	** 999.799 492 2	kg of water (0.00 °C)
ubic metres	** 998.8590	kg of water (15.18 °C)
ubic metres	** 998.681 740 1	kg of water (17.00 °C)
ubic metres	** 999.940 003 6	kg of water (4.00 °C)
ubic metres	* 1000.0	litres
ubic metres	1759.753 986	pints (UK)
ubic metres	2113.376 418	pints (US liquid)
ubic metres	1056.688 209	quarts (US liquid)
ubic metres	1.0	steres
ubic millimetres	0.0010	cubic centimetres
ubic millimetres	6.1024×10^{-5}	cubic inches
ubic millimetres	1.0000×10^{-9}	cubic metres
ubic millimetres	0.016 893 638	minims (UK)
ubic millimetres	0.016 230 731	minims (US)
ubic yards	764 554.8580	cubic centimetres
ubic yards	764.554 858 0	cubic decimetres

Volume (Capacity)

MULTIPLY	BY	TO OBTAIN
cubic yards	27.0	cubic feet
cubic yards	46 656.0	cubic inches
cubic yards	0.764 554 858	cubic metres
cubic yards	168.178 557 4	gallons (UK)
cubic yards	173.569 814 4	gallons (US dry)
cubic yards	201.974 025 9	gallons (US liquid)
cubic yards	764.554 858 2	litres
cubic yards	26 908.569 18	ounces (UK fluid)
cubic yards	25 852.675 33	ounces (US fluid)
cubic yards	1345.428 459	pints (UK)
cubic yards	1615.792 207	pints (US)
cubic yards	672.714 229 5	quarts (UK)
cubic yards	694.279 257 6	quarts (US dry)
cubic yards	807.896 103 8	quarts (US liquid)
decilitres	0.10	litres
dekalitres	10.0	litres
diameter (sphere) cubed	0.523 598 776	volume (sphere)
drachms (UK fluid)	3.551 632 813	cubic centimetres
drachms (UK fluid)	0.216 733 932	cubic inches
drachms (UK fluid)	0.960 759 940	drams (US fluid)
drachms (UK fluid)	3.551 632 813	millilitres
drams (US fluid)	3.696 911 96	cubic centimetres
drams (US fluid)	0.225 585 938	cubic inches
drams (US fluid)	1.040 842 731	drachms (UK fluid)
drams (US fluid)	0.031 250	gills (US)
drams (US fluid)	3.696 691 196	millilitres
drams (US fluid)	* 60.0	minims (US)
drams (US fluid)	0.1250	ounces (US fluid)
drams (US fluid)	0.007 812 50	pints (US liquid)
firkins (UK)	1.1250	bushels (UK)
firkins (UK)	40 914.810	cubic centimetres
firkins (UK)	1.444 892 879	cubic feet
firkins (UK)	1.200 949 926	firkins (US)
firkins (UK)	* 9.0	gallons (UK)
firkins (UK)	40.914 810	litres
firkins (UK)	72.0	pints (UK)
firkins (US)	0.294 642 857	barrels (US dry)
firkins (US)	0.285 714 286	barrels (US liquid)
firkins (US)	0.966 787 884	bushels (US)
firkins (US)	1.203 125 001	cubic feet
firkins (US)	0.832 674 185	firkins (UK)

Volume (Capacity)

MULTIPLY	BY		TO OBTAIN
firkins (US)	*	9.0	gallons (US liquid)
firkins (US)		34.068 706 07	litres
firkins (US)		72.0	pints (US liquid)
gallons (UK)		3.6856×10^{-6}	acre feet
gallons (UK)		1/36	barrels (UK beer)
gallons (UK)		0.031 746 032	barrels (wine)
gallons (UK)		0.1250	bushels (UK)
gallons (UK)		4546.090	cubic centimetres
gallons (UK)		4.546 090	cubic decimetres
gallons (UK)		0.160 543 653	cubic feet
gallons (UK)		277.419 432 7	cubic inches
gallons (UK)		0.004 546 090	cubic metres
gallons (UK)		0.005 946 061	cubic yards
gallons (UK)		1280.0	drachms (UK fluid)
gallons (UK)	*	1/9	firkins (UK)
gallons (UK)		1.200 949 925	gallons (US liquid)
gallons (UK)		32.0	gills (UK)
gallons (UK)	*	4.546 090	litres
gallons (UK)		4546.090	millilitres
gallons (UK)		76 800.0	minims (UK)
gallons (UK)	*	160.0	ounces (UK fluid)
gallons (UK)		153.721 590 4	ounces (US fluid)
gallons (UK)		0.50	pecks (UK)
gallons (UK)		8.0	pints (UK)
gallons (UK)		9.607 599 40	pints (US)
gallons (UK)	*	2.0	pottles
gallons (UK)		10.010 977 18	pounds water (15.18 °C)
gallons (UK)		10.009 200 61	pounds water (17.00 °C)
gallons (UK)		10.021 811 45	pounds water (4.00 °C)
gallons (UK)	*	4.0	quarts (UK)
gallons (US dry)		0.038 095 592	barrels (US dry)
gallons (US dry)		0.036 941 180	barrels (US liquid)
gallons (US dry)		0.1250	bushels (US)
gallons (US dry)		4404.883 769	cubic centimetres
gallons (US dry)		0.155 557 002	cubic feet
gallons (US dry)		268.802 499 8	cubic inches
gallons (US dry)		0.968 938 972	gallons (UK)
gallons (US dry)		1.163 647 185	gallons (US liquid)
gallons (US dry)		4.404 883 769	litres
gallons (US liquid)		3.0689×10^{-6}	acre feet
gallons (US liquid)		0.031 746 032	barrels (US liquid)

Volume (Capacity)

MULTIPLY	BY		TO OBTAIN
gallons (US liquid)		0.023 809 524	barrels (US oil)
gallons (US liquid)		0.107 420 876	bushels (US)
gallons (US liquid)		3785.411 785	cubic centimetres
gallons (US liquid)		3.785 411 785	cubic decimetres
gallons (US liquid)		0.133 680 556	cubic feet
gallons (US liquid)	*	231.0	cubic inches
gallons (US liquid)		0.003 785 412	cubic metres
gallons (US liquid)		0.004 951 132	cubic yards
gallons (US liquid)	**	0.832 674 185	gallons (UK)
gallons (US liquid)		0.859 367 008	gallons (US dry)
gallons (US liquid)		32.0	gills (US)
gallons (US liquid)		3.785 411 785	litres
gallons (US liquid)		0.003 785 412	metres cubed
gallons (US liquid)		3785.411 785	millilitres
gallons (US liquid)		61 440.0	minims (US)
gallons (US liquid)		133.227 869 6	ounces (UK fluid)
gallons (US liquid)	*	128.0	ounces (US fluid)
gallons (US liquid)		6.661 393 480	pints (UK)
gallons (US liquid)		8.0	pints (US liquid)
gallons (US liquid)		8.335 882 267	pounds water (15.18 °C)
gallons (US liquid)		8.334 402 961	pounds water (17.00 °C)
gallons (US liquid)		8.344 903 680	pounds water (4.00 °C)
gallons (US liquid)	*	4.0	quarts (US liquid)
gills (UK)		142.065 312 5	cubic centimetres
gills (UK)		8.669 357 275	cubic inches
gills (UK)		0.031 250	gallons (UK)
gills (UK)		1.200 949 925	gills (US)
gills (UK)		0.142 065 313	litres
gills (UK)	*	5.0	ounces (UK fluid)
gills (UK)		4.803 799 700	ounces (US fluid)
gills (UK)		0.250	pints (UK)
gills (US)		118.294 118 3	cubic centimetres
gills (US)		7.218 750 003	cubic inches
gills (US)		32.0	drams (US fluid)
gills (US)		0.031 25	gallons (US liquid)
gills (US)		0.832 674 185	gills (UK)
gills (US)		0.118 294 118	litres
gills (US)		1920.0	minims (US)
gills (US)	*	4.0	ounces (US fluid)
gills (US)		0.250	pints (US liquid)
gills (US)		0.1250	quarts (US liquid)

Volume (Capacity)

MULTIPLY	BY	TO OBTAIN
hectolitres	2.749 615 604	bushels (UK)
hectolitres	2.837 759 327	bushels (US)
hectolitres	100 000.0	cubic centimetres
hectolitres	3.531 466 672	cubic feet
hectolitres	0.10	cubic metres
hectolitres	21.996 924 83	gallons (UK)
hectolitres	26.417 205 23	gallons (US liquid)
hectolitres	100.0	litres
hectolitres	3519.507 973	ounces (UK fluid)
hectolitres	3381.402 269	ounces (US fluid)
hectolitres	11.351 037 31	pecks (US)
hogsheads	0.5	butts (UK)
hogsheads	8.428 541 795	cubic feet
hogsheads	14 564.520 22	cubic inches
hogsheads	0.238 669 725	cubic metres
hogsheads	* 52.5	gallons (UK)
hogsheads	63.049 871 07	gallons (US liquid)
hogsheads	238.669 725 0	litres
kilderkins (UK)	81 829.620	cubic centimetres
kilderkins (UK)	2.889 785 758	cubic feet
kilderkins (UK)	4993.549 790	cubic inches
kilderkins (UK)	0.081 829 620	cubic metres
kilderkins (UK)	* 18.0	gallons (UK)
kilolitres	1.0×10^6	cubic centimetres
kilolitres	35.314 666 72	cubic feet
kilolitres	61 023.744 09	cubic inches
kilolitres	1.0	cubic metres
kilolitres	1.307 950 619	cubic yards
kilolitres	219.969 248 3	gallons (UK)
kilolitres	227.020 746 2	gallons (US dry)
kilolitres	264.172 052 3	gallons (US liquid)
kilolitres	1000.0	litres
litres	8.1071×10^{-7}	acre feet
litres	0.027 496 156	bushels (UK)
litres	0.028 377 593	bushels (US)
litres	1000.0	cubic centimetres
litres	* 1.0	cubic decimetres
litres	0.035 314 667	cubic feet
litres	61.023 744 09	cubic inches
litres	0.0010	cubic metres
litres	0.001 307 951	cubic yards

Volume (Capacity)

MULTIPLY	BY	TO OBTAIN
litres	270.512 181 6	drams (US fluid)
litres	0.219 969 248	gallons (UK)
litres	0.227 020 746	gallons (US dry)
litres	0.264 172 052	gallons (US liquid)
litres	7.039 015 946	gills (UK)
litres	8.453 505 672	gills (US)
litres	0.004 189 890	hogsheads
litres	16 893.638 27	minims (UK)
litres	16 320.730 89	minims (US)
litres	35.195 079 73	ounces (UK fluid)
litres	33.814 022 69	ounces (US fluid)
litres	0.109 984 624	pecks (UK)
litres	0.113 510 373	pecks (US)
litres	1.759 753 986	pints (UK)
litres	1.816 165 969	pints (US dry)
litres	2.113 376 418	pints (US liquid)
litres	2.202 107 126	pounds water (15.18 °C)
litres	2.201 716 334	pounds water (17.00 °C)
litres	2.204 490 331	pounds water (4.00 °C)
litres	0.879 876 993	quarts (UK)
litres	0.908 082 985	quarts (US dry)
litres	1.056 688 209	quarts (US liquid)
microlitres	1.0×10^{-6}	litres
mil feet	9.4248×10^{-6}	cubic inches
millilitres	1.0	cubic centimetres
millilitres	0.061 023 744	cubic inches
millilitres	0.270 512 182	drams (US fluid)
millilitres	0.000 845 350 6	gills (US)
millilitres	0.0010	litres
millilitres	16.230 730 89	minims (US)
millilitres	0.035 195 080	ounces (UK fluid)
millilitres	0.033 814 023	ounces (US fluid)
millilitres	0.001 759 754	pints (UK)
millilitres	0.002 113 376	pints (US liquid)
minims (UK)	0.059 193 905	cubic centimetres
minims (UK)	0.003 612 234	cubic inches
minims (UK)	0.059 193 905	millilitres
minims (UK)	1/480	ounces (UK fluid)
minims (UK)	0.050	scruples (UK fluid)
minims (US fluid)	0.061 611 520	cubic centimetres
minims (US fluid)	0.003 759 766	cubic inches

Volume (Capacity)

MULTIPLY	BY	TO OBTAIN
minims (US fluid)	1/60	drams (US fluid)
minims (US fluid)	1.6276×10^{-5}	gallons (US liquid)
minims (US fluid)	0.000 520 833	gills (US)
minims (US fluid)	6.1612×10^{-5}	litres
minims (US fluid)	0.061 611 520	millilitres
minims (US fluid)	0.002 083 333	ounces (US fluid)
minims (US fluid)	0.000 130 208	pints (US liquid)
noggins (UK)	142.065 312 5	cubic centimetres
noggins (UK)	0.031 250	gallons (UK)
noggins (UK)	* 1.0	gills (UK)
ounces (UK fluid)	28.413 062 50	cubic centimetres
ounces (UK fluid)	0.028 413 063	cubic decimetres
ounces (UK fluid)	0.001 003 398	cubic feet
ounces (UK fluid)	1.733 871 455	cubic inches
ounces (UK fluid)	3.7163×10^{-5}	cubic yards
ounces (UK fluid)	* 8.0	drachms (UK fluid)
ounces (UK fluid)	7.686 079 521	drams (US fluid)
ounces (UK fluid)	0.006 250	gallons (UK)
ounces (UK fluid)	0.007 505 937	gallons (US liquid)
ounces (UK fluid)	0.028 413 063	litres
ounces (UK fluid)	28.413 062 50	millilitres
ounces (UK fluid)	* 480.0	minims (UK)
ounces (UK fluid)	0.960 759 940	ounces (US fluid)
ounces (UK fluid)	0.050	pints (UK)
ounces (US fluid)	29.573 529 57	cubic centimetres
ounces (US fluid)	0.029 573 530	cubic decimetres
ounces (US fluid)	0.001 044 379	cubic feet
ounces (US fluid)	1.804 687 50	cubic inches
ounces (US fluid)	2.9574×10^{-5}	cubic metres
ounces (US fluid)	3.8681×10^{-5}	cubic yards
ounces (US fluid)	* 8.0	drams (US fluid)
ounces (US fluid)	0.006 505 267	gallons (UK)
ounces (US fluid)	0.006 713 805	gallons (US dry)
ounces (US fluid)	0.007 812 50	gallons (US liquid)
ounces (US fluid)	0.250	gills (US)
ounces (US fluid)	0.029 573 530	litres
ounces (US fluid)	29.573 529 57	millilitres
ounces (US fluid)	* 480.0	minims (US)
ounces (US fluid)	1.040 842 731	ounces (UK fluid)
ounces (US fluid)	0.052 042 137	pints (UK)
ounces (US fluid)	0.062 50	pints (US liquid)

Volume (Capacity)

MULTIPLY	BY	TO OBTAIN
ounces (US fluid)	0.031 250	quarts (US liquid)
pecks (UK)	0.250	bushels (UK)
pecks (UK)	9092.180	cubic centimetres
pecks (UK)	554.838 865 6	cubic inches
pecks (UK)	* 2.0	gallons (UK)
pecks (UK)	64.0	gills (UK)
pecks (UK)	0.038 095 238	hogsheads
pecks (UK)	1/9	kilderkins (UK)
pecks (UK)	9.092 180	litres
pecks (UK)	1.032 056 744	pecks (US)
pecks (UK)	16.0	pints (UK)
pecks (UK)	* 4.0	quarterns (UK dry)
pecks (UK)	8.0	quarts (UK)
pecks (UK)	8.256 453 952	quarts (US dry)
pecks (US)	0.076 191 185	barrels (US dry)
pecks (US)	0.250	bushels (US)
pecks (US)	8809.767 538	cubic centimetres
pecks (US)	0.311 114 005	cubic feet
pecks (US)	537.604 999 8	cubic inches
pecks (US)	* 2.0	gallons (US dry)
pecks (US)	2.327 294 371	gallons (US liquid)
pecks (US)	8.809 767 538	litres
pecks (US)	* 16.0	pints (US dry)
pecks (US)	* 8.0	quarts (US dry)
pints (UK)	568.261 250	cubic centimetres
pints (UK)	0.568 261 250	cubic decimetres
pints (UK)	0.020 067 957	cubic feet
pints (UK)	34.677 429 10	cubic inches
pints (UK)	0.000 568 261	cubic metres
pints (UK)	0.000 743 258	cubic yards
pints (UK)	0.1250	gallons (UK)
pints (UK)	0.150 118 741	gallons (US)
pints (UK)	* 4.0	gills (UK)
pints (UK)	4.803 799 699	gills (US)
pints (UK)	0.568 261 250	litres
pints (UK)	9600.0	minims (UK)
pints (UK)	* 20.0	ounces (UK fluid)
pints (UK)	19.215 198 80	ounces (US fluid)
pints (UK)	1.032 056 744	pints (US dry)
pints (UK)	1.200 949 925	pints (US liquid)
pints (UK)	* 4.0	quarterns (UK liquid)

Volume (Capacity)

MULTIPLY	BY	TO OBTAIN
pints (UK)	0.5	quarts (UK)
pints (US dry)	550.610 471 1	cubic centimetres
pints (US dry)	0.019 444 625	cubic feet
pints (US dry)	33.600 312 48	cubic inches
pints (US dry)	0.000 550 610	cubic metres
pints (US dry)	0.000 720 171	cubic yards
pints (US liquid)	473.176 473 1	cubic centimetres
pints (US liquid)	0.016 710 069	cubic feet
pints (US liquid)	28.875 000 01	cubic inches
pints (US liquid)	0.000 473 176	cubic metres
pints (US liquid)	0.000 618 891	cubic yards
pints (US liquid)	* 128.0	drams (US fluid)
pints (US liquid)	0.1250	gallons (US liquid)
pints (US liquid)	3.330 696 739	gills (UK)
pints (US liquid)	* 4.0	gills (US)
pints (US liquid)	0.473 176 473	litres
pints (US liquid)	473.176 473 1	millilitres
pints (US liquid)	7680.0	minims (US)
pints (US liquid)	* 16.0	ounces (US fluid)
pints (US liquid)	0.832 674 185	pints (UK)
pints (US liquid)	0.5	quarts (US liquid)
pottles	2273.0450	cubic centimetres
pottles	0.002 273 045	cubic metres
pottles	0.50	gallons (UK)
pottles	2.273 045 0	litres
puncheons (UK)	0.317 974 590	cubic metres
puncheons (UK)	69.944 631 52	gallons (UK)
puncheons (UK)	* 84.0	gallons (US liquid)
quarterns (UK dry)	0.1250	buckets (UK)
quarterns (UK dry)	0.062 50	bushels (UK)
quarterns (UK dry)	2273.0450	cubic centimetres
quarterns (UK dry)	0.50	gallons (UK)
quarterns (UK dry)	2.273 045 0	litres
quarterns (UK dry)	0.250	pecks (UK)
quarterns (UK liquid)	142.065 312 5	cubic centimetres
quarterns (UK liquid)	0.031 250	gallons (UK)
quarterns (UK liquid)	0.142 065 313	litres
quarterns (UK liquid)	0.250	pints (UK)
quarts (UK)	1136.522 50	cubic centimetres
quarts (UK)	69.354 858 20	cubic inches
quarts (UK)	0.250	gallons (UK)

Volume (Capacity)

MULTIPLY	BY		TO OBTAIN
quarts (UK)		0.300 237 481	gallons (US liquid)
quarts (UK)		1.136 522 50	litres
quarts (UK)		1.032 056 744	quarts (US dry)
quarts (UK)		1.200 949 925	quarts (US liquid)
quarts (US dry)		0.031 250	bushels (US)
quarts (US dry)		1101.220 942	cubic centimetres
quarts (US dry)		0.038 889 251	cubic feet
quarts (US dry)		67.200 624 96	cubic inches
quarts (US dry)		0.250	gallons (US dry)
quarts (US dry)		0.290 911 796	gallons (US liquid)
quarts (US dry)		1.101 122 094 2	litres
quarts (US dry)		0.1250	pecks (US)
quarts (US dry)	*	2.0	pints (US dry)
quarts (US liquid)		946.352 946 3	cubic centimetres
quarts (US liquid)		0.033 420 14	cubic feet
quarts (US liquid)		57.750 000 02	cubic inches
quarts (US liquid)		0.000 946 353	cubic metres
quarts (US liquid)		0.001 237 783	cubic yards
quarts (US liquid)		256.0	drams (US fluid)
quarts (US liquid)		0.214 841 752	gallons (US dry)
quarts (US liquid)		0.250	gallons (US liquid)
quarts (US liquid)	*	8.0	gills (US)
quarts (US liquid)		0.946 352 946	litres
quarts (US liquid)	*	32.0	ounces (US fluid)
quarts (US liquid)	*	2.0	pints (US liquid)
quarts (US liquid)		0.832 674 185	quarts (UK)
quarts (US liquid)		0.859 367 008	quarts (US dry)
scruples (UK fluid)	*	20.0	minims (UK)
seams (UK)	*	8.0	bushels (UK)
seams (UK)		10.274 793 81	cubic feet
seams (UK)		0.290 949 760	cubic metres
seams (UK)		290.949 760 1	litres
steres	*	1.0	cubic metres
steres		0.1	decasteres
steres		10.0	decisteres
steres		1000.0	litres
tuns		210.0	gallons (UK)
tuns		252.199 484 3	gallons (US liquid)
tuns (brewers')	*	4.0	hogsheads

VOLUME FLOW
Dimensions: L^3/T

MULTIPLY	BY	TO OBTAIN
acre feet/hour	726.0	cubic feet/minute
acre feet/hour	4522.134 545	gallons (UK)/minute
acre feet/hour	5430.857 143	gallons (US)/minute
cubic centimetres/second	0.002 118 880	cubic feet/minute
cubic centimetres/second	1.0000×10^{-6}	cubic metres/second
cubic centimetres/second	0.013 198 155	gallons (UK)/minute
cubic centimetres/second	0.000 219 969	gallons (UK)/second
cubic centimetres/second	0.015 850 323	gallons (US)/minute
cubic centimetres/second	0.000 264 172	gallons (US)/second
cubic centimetres/second	3.60	litres/hour
cubic feet/hour	2.2957×10^{-5}	acre feet/hour
cubic feet/hour	7.865 790 719	cubic centimetres/s
cubic feet/hour	0.028 316 847	cubic metres/hour
cubic feet/hour	7.8658×10^{-6}	cubic metres/second
cubic feet/hour	6.228 835 459	gallons (UK)/hour
cubic feet/hour	7.480 519 478	gallons (US)/hour
cubic feet/hour	0.471 947 443	litres/minute
cubic feet/minute	0.033 057 851	acre feet/24 hours
cubic feet/minute	0.001 377 411	acre feet/hour
cubic feet/minute	2.2957×10^{-5}	acre feet/minute
cubic feet/minute	471.947 443 2	cubic centimetre/s
cubic feet/minute	60.0	cubic feet/hour
cubic feet/minute	1.699 010 795	cubic metres/hour
cubic feet/minute	0.028 316 847	cubic metres/minute
cubic feet/minute	0.000 471 947	cubic metres/second
cubic feet/minute	1.0	cumins
cubic feet/minute	6.228 835 460	gallons (UK)/minute
cubic feet/minute	0.103 813 924	gallons (UK)/second
cubic feet/minute	8969.523 060	gallons (UK)/24 hours
cubic feet/minute	10 771.948 05	gallons (US)/24 hours
cubic feet/minute	7.480 519 480	gallons (US)/minute
cubic feet/minute	0.124 675 325	gallons (US)/second
cubic feet/minute	62.345 663 67	lb water (17 °C)/min
cubic feet/minute	62.424 214 51	lb water (4 °C)/min
cubic feet/minute	1699.010 795	litres/hour
cubic feet/minute	0.471 947 443	litres/second
cubic feet/second	1.983 471 074	acre feet/24 hours
cubic feet/second	0.991 735 538	acre inches/hour

Volume Flow

MULTIPLY	BY	TO OBTAIN
cubic feet/second	28 316.846 59	cubic centimetres/s
cubic feet/second	0.028 316 847	cubic metres/second
cubic feet/second	2.222 222 222	cubic yards/minute
cubic feet/second	1.0	cusecs
cubic feet/second	$5.381\ 71 \times 10^5$	gallons (UK)/24 hours
cubic feet/second	373.730 127 5	gallons (UK)/minute
cubic feet/second	$6.463\ 17 \times 10^5$	gallons (US)/24 hours
cubic feet/second	448.831 168 7	gallons (US)/minute
cubic feet/second	1699.010 795	litres/minute
cubic feet/second	28.316 846 58	litres/second
cubic feet/second	0.646 316 883	millions US gal/day
cubic metres/day	1.1574×10^{-5}	cubic metres/second
cubic metres/hour	0.588 577 779	cubic feet/minute
cubic metres/hour	1/3600	cubic metres/second
cubic metres/minute	35.314 666 72	cubic feet/minute
cubic metres/minute	219.969 248 3	gallons (UK)/minute
cubic metres/minute	264.172 052 3	gallons (US)/minute
cubic metres/minute	1000.0	litres/minute
cubic yards/minute	0.450	cubic feet/second
cubic yards/minute	3.366 233 765	gallons (US)/second
cubic yards/minute	12.742 580 97	litres/second
gallons (UK)/hour	1.262 802 778	cubic centimetres/s
gallons (UK)/hour	0.160 543 653	cubic feet/hour
gallons (UK)/hour	0.002 675 728	cubic feet/minute
gallons (UK/hour	4.4595×10^{-5}	cubic feet/second
gallons (UK)/hour	0.004 546 090	cubic metres/hour
gallons (UK)/hour	7.5768×10^{-5}	cubic metres/minute
gallons (UK)/hour	1.2628×10^{-6}	cubic metres/second
gallons (UK)/hour	4.546 089 997	litres/hour
gallons (UK)/minute	0.272 765 40	cubic metres/hour
gallons (UK)/minute	0.004 546 090	cubic metres/minute
gallons (UK)/minute	7.5768×10^{-5}	cubic metres/second
gallons (UK)/second	4546.089 997	cubic centimetres/s
gallons (US)/day	0.005 570 023	cubic feet/hour
gallons (US)/day	9.2834×10^{-5}	cubic feet/minute
gallons (US)/day	1.5472×10^{-6}	cubic feet/second
gallons (US)/day	0.000 157 725	cubic metres/hour
gallons (US)/day	2.6288×10^{-6}	cubic metres/minute
gallons (US)/day	4.3813×10^{-8}	cubic metres/second
gallons (US)/hour	3.0689×10^{-6}	acre feet/hour
gallons (US)/hour	1.051 503 274	cubic centimetres/s

Volume Flow

MULTIPLY	BY	TO OBTAIN
gallons (US)/hour	0.133 680 556	cubic feet/hour
gallons (US)/hour	0.003 785 412	cubic metres/hour
gallons (US)/hour	6.3090×10^{-5}	cubic metres/minute
gallons (US)/hour	1.0515×10^{-6}	cubic metres/second
gallons (US)/hour	8.2519×10^{-5}	cubic yards/minute
gallons (US)/hour	3.785 411 785	litres/hour
gallons (US)/minute	0.004 419 192	acre feet/24 hours
gallons (US)/minute	8.020 833 336	cubic feet/hour
gallons (US)/minute	0.133 680 556	cubic feet/minute
gallons (US)/minute	0.002 228 009	cubic feet/second
gallons (US)/minute	0.227 124 707	cubic metres/hour
gallons (US)/minute	0.003 785 412	cubic metres/minute
gallons (US)/minute	6.3090×10^{-5}	cubic metres/second
gallons (US)/minute	3.785 411 785	litres/minute
gallons (US)/minute	0.063 090 196	litres/second
gallons (US)/second	3785.411 785	cubic centimetres/s
gallons (US)/second	8.020 833 333	cubic feet/minute
gallons (US)/second	0.297 067 901	cubic yards/minute
gallons (US)/second	227.124 707 1	litres/minute
lb water (17 °C)/min	0.000 267 327	cubic feet/second
lb water (17 °C)/min	0.099 908 079	gallons (UK)/minute
lb water (17 °C)/min	0.119 984 599	gallons (US liquid)/min
lb water (17 °C)/min	0.454 191 117	litres/minute
lb water (4 °C)/min	0.000 266 990	cubic feet/second
lb water (4 °C)/min	0.099 782 360	gallons (UK)/minute
lb water (4 °C)/min	0.119 833 618	gallons (US)/minute
lb water (4 °C)/min	0.453 619 590	litres/minute
litres/hour	0.0010	cubic metres/hour
litres/hour	1/60 000	cubic metres/minute
litres/hour	2.7778×10^{-7}	cubic metres/second
litres/minute	0.035 314 667	cubic feet/minute
litres/minute	0.000 588 578	cubic feet/second
litres/minute	0.060	cubic metres/hour
litres/minute	1/60 000	cubic metres/second
litres/minute	0.219 969 248	gallons (UK)/minute
litres/minute	0.003 666 154	gallons (UK)/second
litres/minute	0.264 172 052	gallons (US liquid)/min
litres/minute	0.004 402 868	gallons (US liquid)/s
litres/second	2.118 880 003	cubic feet/minute
litres/second	0.035 314 667	cubic feet/second
litres/second	3.60	cubic metres/hour

Volume Flow

MULTIPLY	BY	TO OBTAIN
litres/second	0.0010	cubic metres/second
litres/second	0.078 477 037	cubic yards/minute
litres/second	15.850 323 14	gallons (US liquid)/min
litres/second	0.264 172 052	gallons (US liquid)/s
millions UK gallons/day	4546.090	cubic metres/day
millions UK gallons/day	0.052 616 782	cubic metres/second
millions US gallons/day	1.547 228 652	cubic feet/second
miners' inches	1.5	cubic feet/minute

Primary conversion factors

ACCELERATION (Linear)
Dimensions: L/T^2

MULTIPLY	BY		TO OBTAIN
accel'n by gravity,(g)	* 9.806 650	†	metres/second²

AREA
Dimensions: L^2

MULTIPLY	BY	TO OBTAIN
acres	* 4.0	roods (UK)
acres	* 4840.0	square yards
ares	* 100.0	square metres
centares (centiares)	* 1.0	square metres
roods (UK)	* 40.0	square perches
square feet (US survey)	** 1.000 004 0	square feet

CIRCULAR and SPHERICAL GEOMETRY
Dimensions: Various

MULTIPLY	BY	TO OBTAIN
circles	* 1.0	revolutions
circumferences	* 1.0	revolutions
degrees	* 60.0	minutes
minutes	* 60.0	seconds
quadrants	* 90.0	degrees
radians	** 57.295 779 51	degrees
revolutions	* 360.0	degrees
revolutions	* 400.0	grades
revolutions	* 4.0	quadrants
spheres	* 1.0	solid angles
spheres	* 8.0	spherical right angles
spheres	** 12.566 370 62	steradians

* denotes an exact number; ** denotes an inexact number. See the Introduction for further discussion.

† local gravity varies

ELECTRICAL (Charge)
Dimensions: C

MULTIPLY	BY		TO OBTAIN
CGS e.m. units	*	10.0	coulombs
CGS e.s. units	*	1.0	Statcoulombs
coulombs	*	1.0	ampere seconds
faradays (chem)	**	26.802 773 29	ampere hours
faradays (phys)	**	26.810 301 25	ampere hours
Statcoulombs	**	333.564 604 8	picocoulombs

ELECTRICAL (Conductance)
Dimensions: C^2T/ML^2

MULTIPLY	BY		TO OBTAIN
CGS e.m. units	*	1.0×10^9	siemens
CGS e.s. units	*	1.0	Statmhos
mhos	*	1.0	siemens
mhos (International)	**	0.999 505 245	siemens
Statmhos	**	1.112 653 456	picosiemens units

ELECTRICAL (Conductivity)
Dimensions: C^2T/ML^3

MULTIPLY	BY		TO OBTAIN
CGS e.m. units	*	1.0×10^{11}	siemens/metre
CGS e.s. units	*	1.0	Statmhos/centimetre
Statmhos/centimetre	**	111.265 345 6	picosiemens/metre

ELECTRICAL (Current)
Dimensions: C/T

MULTIPLY	BY		TO OBTAIN
Abamperes	*	10.0	amperes
Abamperes	*	10.0	coulombs/second
amperes (International)	**	0.999 835 027	amperes
CGS e.m. units	*	10.0	amperes
CGS e.s. units	*	1.0	Statamperes
coulombs/second	*	1.0	amperes
faradays (chem)/second	**	9648.998 385	Abamperes
faradays (phys)/second	**	9651.708 450	Abamperes
Statamperes	**	333.564 604 8	picoamperes

ELECTRICAL (Electric capacitance)
Dimensions: C^2T^2/ML^2

MULTIPLY	BY		TO OBTAIN
Abfarads	*	1.0×10^9	farads
CGS e.m. units	*	1.0×10^9	farads
CGS e.s. units	*	1.0	Statfarads
farads (International)	**	0.999 505 245	farads
Statfarads	**	1.112 653 456	picofarads

ELECTRICAL (Electric field)
Dimensions: ML/CT^2

MULTIPLY	BY		TO OBTAIN
CGS e.m. units	*	1.0×10^{-6}	volts/metre
CGS e.s. units	*	1.0	Statvolts/centimetre

ELECTRICAL (Electric potential)
Dimensions: ML^2/CT^2

MULTIPLY	BY		TO OBTAIN
Abvolts	*	1.0×10^{-8}	volts
CGS e.m. units	*	1.0×10^{-8}	volts
CGS e.s. units	*	1.0	Statvolts
CGS e.s. units	**	299.7920	volts
volts (International)	**	1.000 331 0	volts

ELECTRICAL (Inductance)
Dimensions: ML^2/C^2

MULTIPLY	BY		TO OBTAIN
Abhenries	*	1.0×10^{-9}	henries
CGS e.m. units	*	1.0×10^{-9}	henries
CGS e.s. units	*	1.0	Stathenries
Stathenries	**	898.752 432 4	gigahenries

ELECTRICAL (Linear current density)
Dimensions: C/LT

MULTIPLY	BY		TO OBTAIN
CGS e.m. units	*	1.0	oersteds
CGS e.s. units	**	2654.422 783	picoamperes/metre

ELECTRICAL (Magnetic field strength)
Dimensions: C/LT

MULTIPLY	BY	TO OBTAIN
CGS e.m. units	** 79.577 471 51	amperes/metre
CGS e.m. units	* 1.0	oersteds
CGS e.s. units	** 2654.422 783	picoamperes/metre
oersteds (International)	** 0.999 835 027	oersteds

ELECTRICAL (Magnetic flux density)
Dimensions: M/CT

MULTIPLY	BY	TO OBTAIN
CGS e.m. units	* 1.0	gausses
CGS e.s. units	** $2.997\ 92 \times 10^6$	teslas
gausses (International)	** 1.000 331 0	gausses
maxwells/sq centimetre	* 1.0	gausses
teslas	* 10 000.0	gausses
teslas	* 1.0	webers/sq metre

ELECTRICAL (Magnetic flux)
Dimensions: ML²/CT

MULTIPLY	BY	TO OBTAIN
CGS e.m. units	* 1.0×10^{-8}	webers
CGS e.s. units	** $2.997\ 92 \times 10^2$	webers
lines	* 1.0	maxwells
maxwells	* 1.0×10^{-8}	volt seconds
maxwells	* 1.0×10^{-8}	webers
maxwells (International)	** 1.000 331 0	maxwells

ELECTRICAL (Magnetomotive force)
Dimensions: CL/T

MULTIPLY	BY	TO OBTAIN
CGS e.m. units	** 795.774 715 1	milliampere turns
CGS e.s. units	** 2.6544×10^{-8}	milliampere turns
gilberts	** 795.774 715 1	milliampere turns

ELECTRICAL (Permeability)
Dimensions: ML/C^2

MULTIPLY	BY		TO OBTAIN
CGS e.m. units	**	1.2566×10^{-6}	henries/metre
CGS e.s. units	**	1.1294×10^{15}	henries/metre
gausses/oersted	**	1.256 637 062	microhenries/metre

ELECTRICAL (Resistance)
Dimensions: ML^2/C^2T

MULTIPLY	BY		TO OBTAIN
Abohms	*	1.0×10^{-9}	ohms
CGS e.m. units	*	1.0×10^{-9}	ohms
CGS e.s. units	**	8.8975×10^{11}	ohms

ELECTRICAL (Resistivity)
Dimensions: ML^3/C^2T

MULTIPLY	BY		TO OBTAIN
CGS e.m. units	*	1.0×10^{-11}	ohm metres
CGS e.s. units	**	8987.524 324	megohm metres
circular mil ohms/foot	**	166.242 611 3	Abohm centimetres

ELECTRICAL (Surface current density)
Dimensions: C/TL^2

MULTIPLY	BY		TO OBTAIN
CGS e.m. units	*	1.0×10^5	amperes/square metre

ELECTRICAL (Volume charge density)
Dimensions: C/L^3

MULTIPLY	BY		TO OBTAIN
CGS e.m. units	*	1.0×10^7	coulombs/cubic metre

ENERGY (Heat, work, electrical)
Dimensions: ML^2/T^2

MULTIPLY	BY		TO OBTAIN
calories	*	4.186 80	joules (newton metres)
calories(15)	**	4.185 50	joules (newton metres)
dyne centimetres	*	1.0×10^{-7}	joules (newton metres)

Energy (Heat, work, electrical)

MULTIPLY	BY	TO OBTAIN
ergs	| * 1.0×10^{-7}	| joules (newton metres)
kilograms of ice melted	|** 334.0	| kilojoules (latent heat)
litre atmospheres	| * 101.3250	| joules (newton metres)
thermies	|** 4.185 50 $\times 10^{6}$	| joules (newton metres)
therms	| * 1.0×10^{5}	| Btu

FORCE (Weight)
Dimensions: ML/T^2

MULTIPLY	BY	TO OBTAIN
carats (metric)	| * 200.0	| milligrams
centals	| * 100.0	| pounds (avoirdupois)
drams (apothecary/troy)	| * 60.0	| grains
drams (apothecary/troy)	| * 3.0	| scruples (apothecary)
dynes	| * 1.0×10^{-5}	| newtons (joules/metre)
grains	| * 1.0	| grains (apothecary)
grains	| * 1.0	| grains (avoirdupois)
grains	| * 1.0	| grains (troy)
hundredwts (long) (UK)	| * 112.0	| pounds (avoirdupois)
hundredwts (short) (US)	| * 100.0	| pounds (avoirdupois)
kilograms	| * 9.806 650 †	| newtons (joules/metre)
ounces (apothecary/troy)	| * 8.0	| drams (apothecary/troy)
ounces (apothecary/troy)	| * 480.0	| grains
ounces (apothecary/troy)	| * 20.0	| pennyweights (troy)
ounces (apothecary/troy)	| * 24.0	| scruples (apothecary)
ounces (avoirdupois)	| * 16.0	| drams (avoirdupois)
pennyweights (troy)	| * 24.0	| grains
pounds (apothecary/troy)	| * 5760.0	| grains
pounds (apothecary/troy)	| * 12.0	| ounces (apothecary/troy)
pounds (avoirdupois)	| * 7000.0	| grains
pounds (avoirdupois)	| * 16.0	| ounces (avoirdupois)
pounds (avoirdupois)	|** 32.174 048 56	| poundals
quintals (metric)	| * 100.0	| kilograms
quintals (UK long)	| * 112.0	| pounds (avoirdupois)
quintals (UK short)	| * 100.0	| pounds (avoirdupois)
scruples (apothecary)	| * 20.0	| grains
stones	| * 14.0	| pounds (avoirdupois)
tonnes (metric)	| * 1000.0	| kilograms
tons (long) (UK)	| * 2240.0	| pounds (avoirdupois)
tons (short) (US)	| * 2000.0	| pounds (avoirdupois)

† local gravity varies

HEAT CAPACITY
Dimensions: L^2/T^2

MULTIPLY	BY		TO OBTAIN
Btu/pound °C	*	2.3260	joules/gram °C

ILLUMINATION (Luminous flux)
Dimensions: ML^2/T^3

MULTIPLY	BY		TO OBTAIN
candle power (spherical)	**	12.566 370 62	lumens
lumens	**	0.001 470 588	watts

ILLUMINATION (Luminous incidence)
Dimensions: L^2/MT^3

MULTIPLY	BY		TO OBTAIN
foot candles	*	1.0	lumens/square foot
lux	*	1.0	lumens/square metre
stilbs	**	3.141 592 654	lamberts

ILLUMINATION (Luminous intensity)
Dimensions: ML^2/T^3

MULTIPLY	BY		TO OBTAIN
candles (International)	*	0.950	candles (German)
candles (International)	*	0.960	candles (UK)
carcel units	*	9.610	candles (International)
hefner candles	*	0.90	candles (International)
new candles	*	60.0	candles/sq centimetre

LINEAR MEASURE (Distance or depth)
Dimensions: L

MULTIPLY	BY		TO OBTAIN
angstrom units	*	1.0×10^{-10}	metres
astronomical units	**	1.4960×10^{11}	metres
bolts (US cloth)	*	120.0	linear feet
cables (US)	**	120.0	fathoms
chains (engineers')	*	100.0	feet

Linear Measure (Distance or depth)

MULTIPLY	BY	TO OBTAIN
chains	* 66.0	feet
cubits	* 1.5	feet
dekametres	* 10.0	metres
ells	* 45.0	inches
ems (printers')	* 1/6	inches
fathoms	* 6.0	feet
feet	* 12.0	inches
feet	* 0.304 80	metres
feet (US Survey)	** 1.000 002 0	feet
furlongs	* 660.0	feet
hand spans	* 9.0	inches
hands	* 4.0	inches
inches	* 48.0	irons (footware)
inches	* 12.0	lines (watchmakers')
inches	** 0.025 40	metres
inches	* 1000.0	mils
inches	** 72.270	points (printers') (US,UK)
leagues (nautical, Int)	* 5.5560	kilometres
leagues (nautical, UK)	* 18 240.0	feet
leagues (statute)	* 3.0	miles (statute)
light years	** $9.460\ 55 \times 10^{12}$	kilometres
links (engineers')	* 1.0	feet
links (US Survey)	* 0.660	feet
microns (μ)	* 10 000.0	angstrom units
microns (μ)	* 1.0×10^{-6}	metres
miles (nautical, Int)	* 1852.0	metres
miles (nautical, UK)	* 6080.0	feet
miles (statute)	* 5280.0	feet
miles (statute)	* 8.0	furlongs
myriametres (US)	* 10.0	kilometres
paces	* 2.50	feet
palms	* 3.0	inches
parsecs	** 3.261 60	light years
perches	* 5.50	yards
pica ems (UK, US)	** 0.1660	inches
picas (printers')	* 12.0	points (printers')
points (Didot)	* 0.376	millimetres
points (UK, US)	** 0.355	millimetres
poles	* 1.0	perches
poles	* 1.0	rods (surveyors')
poles	* 5.5	yards

Linear Measure (Distance or depth)

MULTIPLY	BY	TO OBTAIN
rods (surveyors')	* 25.0	links (US Survey)
rods (surveyors')	* 5.50	yards
ropes (UK)	* 20.0	feet
thou (mils)	* 0.025 40	millimetres
yards	* 3.0	feet
yarns/hanks of cotton	* 840.0	yards
yarns/hanks of wool	* 560.0	yards
yarns/leas of linen	* 300.0	yards
yarns/skeins of wool	* 256.0	yards
yarns/spindles of jute	* 14 400.0	yards

MASS
Dimensions: M

MULTIPLY	BY	TO OBTAIN
carats (metric)	* 200.0	milligrams
pounds (avoirdupois)	** 453.592 374 5	grams

POWER
Dimensions: ML^2/T^3

MULTIPLY	BY	TO OBTAIN
cheval-vapeur	* 1.0	horsepower (metric)
horsepower (boiler)	* 34.50	lb water evap/h 100 °C
horsepower (electric)	* 746.0	joules/second (watts)
horsepower (metric)	* 75.0	kilogram metres/second
horsepower (UK)	* 550.0	foot pounds/second
tons of refrigeration	** 3516.90	watts

PRESSURE or STRESS
Dimensions: M/LT^2

MULTIPLY	BY	TO OBTAIN
at (metric atmosphere)	* 1.0	kgf/square centimetre
atmospheres (Standard)	* 1.013 250	bar
atmospheres (Standard)	* 101.3250	kilonewtons/sq metre
atmospheres (Standard)	* 760.0	torrs
bars	* 1.0×10^5	newtons/square metre
baryes	* 1.0	dynes/square centimetre
newtons/square metre	* 10.0	dynes/square centimetre
pascals	* 1.0	newtons/square metre

TIME
Dimensions: T

MULTIPLY	BY		TO OBTAIN
days (mean solar)	*	24.0	hours (mean solar)
days (sidereal)	*	24.0	hours (sidereal)
hours (mean solar)	*	60.0	minutes (mean solar)
hours (sidereal)	*	60.0	minutes (sidereal)
minutes (mean solar)	*	60.0	seconds (mean solar) †
minutes (sidereal)	*	60.0	seconds (sidereal)
months (lunar)	**	29.530 589 04	days (mean solar)
months (mean calendar)	*	730.0	hours (mean solar)
years (calendar)	*	365.0	days (mean solar)
years (calendar)	*	12.0	months (mean calendar)
years (sidereal)	**	365.256 354 2	days (mean solar)
years (tropical)	**	365.242 199 1	days (mean solar)

VELOCITY (Linear)
Dimensions: L/T

MULTIPLY	BY		TO OBTAIN
diam (inches) × RPM	**	4.787 787 204	metres/hr (belt speed)
feet/second	*	0.304 80	metres/second
knots (International)	*	1.8520	kilometres/hour
knots (International)	*	1.0	miles (nautical)/hour

VISCOSITY (Absolute, dynamic)
Dimensions: M/LT

MULTIPLY	BY		TO OBTAIN
centipoise	*	0.010	grams/centimetre second
centipoise	*	1.0	mN seconds/sq metre
poise	*	1.0	dyne second/square cm
poise	*	1.0	grams/centimetre second

VISCOSITY (Kinematic)
Dimensions: L^2/T

MULTIPLY	BY		TO OBTAIN
stokes	*	1.0	sq centimetres/second

† SI unit

VOLUME (Capacity)
Dimensions: L^3

MULTIPLY	BY	TO OBTAIN
bags (UK)	* 3.0	bushels (UK)
barrels (lube-oil, US)	* 42.0	gallons (US liquid)
barrels (UK)	* 1.5	bags (UK)
barrels (UK)	* 4.5	bushels (UK)
barrels (UK)	* 36.0	gallons (UK)
barrels (UK wine)	* 31.50	gallons (UK)
barrels (US dry)	* 7056.0	cubic inches
barrels (US oil)	* 42.0	gallons (US liquid)
barrels (US liquid)	* 7276.50	cubic inches
barrels (US liquid)	* 31.5	gallons (US liquid)
board feet (timber)	* 144.0	cubic inches
buckets (UK)	* 4.0	gallons (UK)
bushels (UK)	* 8.0	gallons (UK)
bushels (US)	* 8.0	gallons (US dry)
bushels (US)	* 4.0	pecks (US)
butts (UK)	* 105.0	gallons (UK)
butts (UK)	* 2.0	hogsheads
cord feet (timber)	* 16.0	cubic feet
cords (timber)	* 128.0	cubic feet
cubic centimetres	* 13.595 10	grams of Hg (0.0 °C)
cubic metres	** 999.799 492 2	kg of water (0.00 °C)
cubic metres	** 998.8590	kg of water (15.18 °C)
cubic metres	** 998.681 740 1	kg of water (17.00 °C)
cubic metres	** 999.940 003 6	kg of water (4.00 °C)
cubic metres	* 1000.0	litres
drams (US fluid)	* 60.0	minims (US)
firkins (UK ale, beer)	* 9.0	gallons (UK)
firkins (US)	* 9.0	gallons (US liquid)
gallons (UK)	* 4.546 090	litres
gallons (UK)	* 160.0	ounces (UK fluid)
gallons (UK)	* 2.0	pottles
gallons (UK)	* 4.0	quarts (UK)
gallons (US liquid)	* 231.0	cubic inches
gallons (US liquid)	** 0.832 674 185	gallons (UK)
gallons (US liquid)	* 128.0	ounces (US fluid)
gallons (US liquid)	* 4.0	quarts (US liquid)
gills (UK)	* 5.0	ounces (UK fluid)
gills (US)	* 4.0	ounces (US fluid)

Volume (Capacity)

MULTIPLY	BY	TO OBTAIN
hogsheads	* 52.5	gallons (UK)
kilderkins (UK beer)	* 18.0	gallons (UK)
litres	* 1.0	cubic decimetres
noggins (UK)	* 1.0	gills (UK)
ounces (UK fluid)	* 8.0	drachms (UK fluid)
ounces (UK fluid)	* 480.0	minims (UK)
ounces (US fluid)	* 8.0	drams (US fluid)
ounces (US fluid)	* 480.0	minims (US)
pecks (UK)	* 2.0	gallons (UK)
pecks (UK)	* 4.0	quarterns (UK dry)
pecks (US)	* 2.0	gallons (US dry)
pecks (US)	* 16.0	pints (US dry)
pecks (US)	* 8.0	quarts (US dry)
pints (UK)	* 4.0	gills (UK)
pints (UK)	* 20.0	ounces (UK fluid)
pints (UK)	* 4.0	quarterns (UK liquid)
pints (US liquid)	* 128.0	drams (US fluid)
pints (US liquid)	* 4.0	gills (US)
pints (US liquid)	* 16.0	ounces (US fluid)
puncheons (UK)	* 84.0	gallons (US liquid)
quarts (US dry)	* 2.0	pints (US dry)
quarts (US liquid)	* 8.0	gills (US)
quarts (US liquid)	* 32.0	ounces (US fluid)
quarts (US liquid)	* 2.0	pints (US liquid)
scruples (UK fluid)	* 20.0	minims (UK)
seams (UK)	* 8.0	bushels (UK)
steres	* 1.0	cubic metres
tuns (brewers')	* 4.0	hogsheads

Numerical data

APPROXIMATE VALUES OF SOME COMMON FLUID PROPERTIES

Water:

Density of water : 1000.0 kilograms/metre^3
Viscosity of water (18 °C) : 0.001 newton seconds/metre^2
Heat capacity of water : 4.0 kilojoules/kilogram K
Thermal conductivity of water : 0.6 watts/metre K
Latent heat of boiling water : 2.0 megajoules/kilogram
Latent heat of fusion of ice : 334.0 kilojoules/kilogram

Air:

Density of air at s.t.p. : 1.0 kilograms/metre^3
Viscosity of air at s.t.p. : 0.000 017 newton seconds/metre^2
Heat capacity of air at s.t.p. : 1.0 kilojoules/kilogram K
Thermal conductivity of air at s.t.p. : 0.024 watts/metre K

APPROXIMATE VALUES OF SOME COMMON GAS PROPERTIES

NB. 1) All gas volumes are from 1 litre of liquid measured at 16 °C and 1 atm.
2) Gas values marked with an asterisk(*) sublime at the temperature given.

Units		g/litre	litres	deg. K	deg. K	deg. K	atm	kJ/litre
Gas	Mol. Wt.	Liquid density	Volume of gas	Boil. point	Crit. temp.	Melt point	Crit. press.	Latent heat
Acetylene	26.0	621	561	*189.6	309.0	—	62.0	—
Air (21% oxygen)	29.0	878	724	78.8	132.5	—	37.2	181.29
Ammonia	17.0	—	—	239.9	405.6	195.5	111.5	—
Argon	39.9	1410	835	87.5	151.0	84.0	48.0	221.90
Carbon dioxide	44.0	1560	836	*194.7	304.3	—	73.0	895.98
Carbon monoxide	28.0	793	672	81.0	134.0	66.0	35.0	171.24
Deuterium	4.0	164	975	23.6	—	18.6	—	50.24
Ethane	30.1	547	429	184.9	305.3	101.2	48.8	267.96
Ethylene	28.0	568	476	169.4	282.9	103.8	50.9	276.33
Fluorine	38.0	1108	693	86.0	172.0	50.0	—	198.87
Freon 115	154.5	—	—	234.5	353.0	167.0	30.8	—
Freon 116	138.0	—	—	195.0	—	172.6	—	—
Freon 12	120.9	—	—	243.4	385.2	115.0	40.8	—

117

Approximate Values of Some Common Gas Properties

Units		g/litre	litres	deg. K	deg. K	deg. K	atm	kJ/litre
Gas	Mol. Wt.	Liquid density	Volume of gas	Boil. point	Crit. temp.	Melt point	Crit. press.	Latent heat
Freon 13	104.4	—	—	191.8	302.1	92.0	38.2	—
Freon 13 B1	148.9	—	—	216.4	340.7	129.9	39.9	—
Freon 14	88.0	—	—	145.2	227.7	89.0	36.9	—
Freon 22	86.5	—	—	232.4	369.2	113.0	48.7	—
Helium (He3)	3.0	—	—	3.2	3.3	—	—	—
Helium (He4)	4.0	125	740	4.2	5.3	—	2.26	2.72
Hydrogen	2.0	71	830	20.4	33.3	14.0	12.8	32.03
Isobutane	58.1	—	—	263.0	407.0	128.0	37.0	—
Ketene	42.0	—	—	232.0	—	122.0	—	—
Krypton	83.7	2155	615	121.4	210.0	104.0	36.9	248.69
Methane	16.0	415	613	111.8	190.7	89.0	45.8	221.90
Methyl Chloride	50.5	—	—	249.1	416.3	175.0	65.8	—
Neon	20.2	1204	1420	27.3	44.5	24.5	25.9	100.48
Nitrogen	28.0	808	682	77.3	126.1	63.3	33.5	161.61
Nitrous oxide	44.0	1226	675	184.0	309.7	182.4	71.7	460.55
Oxygen	32.0	1140	843	90.2	154.4	54.8	49.7	243.67
Propane	44.1	—	—	231.2	368.8	83.3	43.0	—
Propylene	42.1	—	—	226.2	365.5	88.0	45.0	—
Sulphur dioxide	64.1	—	—	263.0	430.4	200.0	77.7	—
Xenon	131.3	3520	636	164.1	289.8	133.0	58.2	349.18

APPROXIMATE VALUES OF SOME COMMON METAL PROPERTIES (Strength)

Units		GN/m²	MN/m²	MN/m²	None
Metal	Condition	Young's modulus	Tensile strength	Yield strength	Poisson' ratio
Aluminium (pure)	wrought/cast	68	65	17	0.34
Aluminium bronze	wrought/cast	106	345	131	—
Aluminium manganese	wrought	67	97	34	—
Antimony	element	78	11	—	—
Barronia	alpha brass	103	494	—	—
Birmabright	cast	71	169	—	—
Bismuth	element	32	—	—	0.33
Brass (70Cu/30Zn)	alpha brass	103	550	448	0.35
Bronze (90Cu/10Sn)	cast	110	260	140	—
Cobalt	element	211	247	—	—
Constantin	element	170	—	—	0.33

Approximate Values of Some Common Metal Properties (Strength)

Units		GN/m²	MN/m²	MN/m²	None
Metal	Condition	Young's modulus	Tensile strength	Yield strength	Poisson's ratio
Copper (pure)	commercially	117	216	69	0.36
Copper	cast/rolled	—	—	—	—
Duralumin	treat/aged	71	525	—	—
Gold	annealed	79	108	—	0.44
Gunmetal (88Cu/10Sn)	wrought/cast	110	262	—	—
Inconel	wrought/cast	220	1003	—	—
Invar (64Fe/36Ni)	annealed	144	463	275	—
Iridium	annealed	524	1003	—	—
Iron (cast)	grey	140	154	—	0.27
Iron (pure)	commercially	200	320	165	—
Iron (wrought)	annealed	193	370	150	0.28
Lead	element	14	16	12	0.44
Magnesium	element	44	154	77	0.29
Mercury	element	—	—	—	—
MG7 Light alloy	wrought	67	324	—	—
Monel (70Ni/30Cu)	wrought/cast	179	540	240	—
Nickel	wrought/cast	206	386	77	0.36
Niobium	element	103	309	—	—
Phosphor bronze	wrought/cast	103	232	—	—
Platinum	element	152	139	—	0.38
Silver	element	76	139	—	0.37
Solder (50Pb/50Sn)	soft	—	45	—	—
Stainless steel 304	softened	201	618	232	—
Stainless steel 310	softened	201	618	232	—
Stainless steel 316	softened	204	618	278	—
Stainless steel 321	softened	201	648	247	—
Stainless steel 347	softened	202	630	266	—
Stainless steel 410	annealed	215	487	261	—
Stainless steel 431	annealed	216	772	587	—
Steel (mild)	plain carbon	207	430	300	0.29
Tantalum	cold worked	186	463	—	—
Tin	ductile	41	97	—	0.36
Titanium (pure)	commercially	110	463	360	0.36
Tungsten	drawn wire	345	3,860	—	—
Tungum	as rolled	110	—	—	—
Vanadium	element	125	818	—	—
Zinc	element	96	140	—	—
Zirconium	annealed	94	309	—	—

APPROXIMATE VALUES OF SOME COMMON TIMBER PROPERTIES (12% MOISTURE)

(Extracted and modified from *The handbook of hardwoods* and other sources)

Units	kg/m³	kN/m²	N/mm²	N/mm²	metres
Timber	Density	Young's modulus	Bending strength	Crushing strength	Impact strength
Ash, European	641	10–12	85–120	35–55	0.9–1.2
Balsa	176	<10	<50	<20	—
Beech, European	737	12–15	85–120	55–85	0.9–1.2
Birch, European	675	12–15	120–175	55–85	0.9–1.2
Cherry	577	10–12	85–120	35–55	0.9–1.2
Chestnut	561	<10	50–85	35–55	<0.6
Dogwood	785	10–12	85–120	55–85	—
Ebony, African	1169	15–20	>175	>85	1.2–1.6
Elm, Dutch	673	<10	<50	20–35	<0.6
Fir, Douglas	481	12–15	120–175	35–55	0.6–0.9
Greenheart	977	>20	>175	>85	1.2–1.6
Hickory	801	15–20	120–175	55–85	>1.6
Lignum-Vitae	1330	—	—	>85	—
Mahogany, African	769	10–12	85–120	55–85	0.6–0.9
Maple, Hard	673	12–15	120–175	55–85	1.2–1.6
Maple, Soft	609	10–12	85–120	35–55	0.6–0.9
Oak, European	801	10–12	85–120	35–55	0.6–0.9
Obeche	368	<10	50–85	20–35	<0.6
Poplar, Italian	416	<10	50–85	35–55	<0.6
Redwood, European	480	10–12	120–175	35–55	0.6–0.9
Rosewood, Indian	880	10–12	120–175	55–85	—
Spruce	420	10–12	85–120	35–55	<0.6
Sycamore	641	<10	85–120	35–55	0.6–0.9
Teak	801	10–12	85–120	55–85	0.6–0.9
Walnut, European	657	—	—	—	—
Willow	529	<10	50–85	20–35	0.6–0.9

APPROXIMATE VALUES OF SOME COMMON METAL PROPERTIES (Thermal)

Units	kg/m³	K	per K	J/(kg K)	W/(m K)
Metal	Density	Melting point	Lin. exp. coeff.	Sp. heat capacity	Thermal conduct'y
Aluminium (pure)	2713	923	0.000 023 7	913.0	230.3
Aluminium Bronze	8250	1263	0.000 017 4	445.8	64.9
Aluminium Manganese	2740	916	0.000 023 2	873.6	173.8
Antimony	6615	904	0.000 011 0	206.4	18.6
Barronia	8775	1303	0.000 018 5	—	108.9
Birmabright	2685	903	0.000 022 9	—	—
Bismuth	9411	544	0.000 013 4	125.6	8.8
Brass (70Cu/30Zn)	8595	1173	0.000 019 4	376.8	108.9
Bronze (90Cu/10Sn)	8913	1323	0.000 019 0	366.6	75.4
Cobalt	8913	1766	0.000 012 5	439.6	69.2
Constantin	8880	1472	0.000 015 8	418.9	25.1
Copper	8930	1356	0.000 016 8	393.1	389.4
Copper (cast/rolled)	8899	1356	0.000 017 3	385.0	357.9
Duralumin	2810	823	0.000 022 6	895.9	129.8
Gold	19 293	1337	0.000 014 4	132.3	301.4
Gunmetal (88Cu/10Sn)	8691	1323	0.000 016 4	376.8	75.4
Inconel	8775	1663	0.000 011 5	456.4	15.0
Invar (64Fe/36Ni)	7999	1679	0.000 001 5	514.9	16.0
Iridium	22 421	2716	0.000 006 5	135.2	146.5
Iron (cast grey)	7058	1423	0.000 013 5	464.7	54.4
Iron (pure)	7806	1808	0.000 012 2	439.6	76.2
Iron (wrought)	7750	1793	0.000 011 1	464.3	27.2
Lead	11 626	600	0.000 029 1	128.9	34.3
Magnesium	1744	923	0.000 026 1	1029.9	159.1
Mercury	13 424	235	0.000 061 0	138.6	8.8
MG7 Light alloy	2630	823	0.000 022 8	962.9	117.2
Monel (70Ni/30Cu)	8830	1573	0.000 013 6	531.7	62.8
Nickel	8885	1728	0.000 013 0	460.6	60.7
Niobium	8553	2741	0.000 007 1	272.1	52.3
Phosphor bronze	8913	1323	0.000 018 0	366.5	75.4
Platinum	21 424	2042	0.000 009 0	135.23	69.1
Silver	10 518	1234	0.000 019 6	232.8	418.7
Solder (50Pb/50Sn)	8996	490	0.000 025 8	182.9	48.6
Stainless steel 304	8027	1713	0.000 017 0	502.4	15.9
Stainless steel 310	7999	1703	0.000 017 0	0.117	13.5

Approximate Values of Some Common Metal Properties (Thermal)

Units	kgs/m³	K	per K	J/(kg K)	W/(m K)
Metal	Density	Melting point	Lin. exp. coeff.	Sp. heat capacity	Thermal conduct'y
Stainless steel 316	7971	1703	0.000 016 0	502.4	16.3
Stainless steel 321	8027	1703	0.000 016 0	502.4	15.9
Stainless steel 347	8027	1703	0.000 017 0	502.4	15.9
Stainless steel 410	7721	1763	0.000 011 0	481.5	24.3
Stainless steel 431	7689	1743	0.000 010 0	628.0	18.8
Steel (Mild)	7806	1773	0.000 015 0	489.9	62.8
Tantalum	16 607	3223	0.000 006 5	152.8	54.4
Tin	7197	504.8	0.000 023 8	234.5	62.4
Titanium	4512	1953	0.000 009 1	527.5	17.0
Tungsten	19 182	3643	0.000 004 4	138.2	166.2
Tungum	8415	1433	0.000 018 2	—	77.0
Vanadium	6117	2008	0.000 008 3	502.4	31.0
Zinc	7197	693	0.000 031 0	389.8	112.2
Zirconium	6477	2123	0.000 005 8	284.7	16.7

DENSITIES OF VARIOUS MATERIALS

Building trade

Units	kg/m³	m³/tonne	lb/ft³	ft³/ton
Material	Metric density	Metric volume	Imperial density	Imperial volume
Ashes	960	1.0420	60	37.38
Ashes (dry, loose)	609	1.6420	38	58.92
Ashes (wet, loose)	753	1.3280	47	47.65
Asphalt (crushed)	721	1.3870	45	49.77
Ballast (with sand)	1600	0.6250	100	22.43
Basalt (solid)	2963	0.3375	185	12.11
Bitumen	1370	0.7300	86	26.19
Bituminous emulsion	1000	1.0000	62	35.88
Brick (fire)	2243	0.4458	140	16.00
Brickwork (ordinary)	1920	0.5210	120	18.69
Brickwork (pressed brick)	2120	0.4720	132	16.93
Cement (aluminous)	1400	0.7140	87	25.63
Cement (bulk)	1281	0.7806	80	28.01
Cement (fast hardening)	1280	0.7810	80	28.03
Cement (Portland)	1440	0.6940	90	24.92
Cement (Portland, aerated)	1041	0.9606	65	34.47
Cement (Portland, clinker)	1794	0.5574	112	20.00

Densities of Various Materials

Building trade

Material	kg/m³ Metric density	m³/tonne Metric volume	lb/ft³ Imperial density	ft³/ton Imperial volume
Units				
Cement (Portland, packed)	1522	0.6570	95	23.58
Chalk (crushed)	1442	0.6935	90	24.88
Chalk (solid)	2483	0.4027	155	14.45
Clay (dry)	1121	0.8921	70	32.01
Clay (wet)	1920	0.5210	120	18.69
Clinker	800	0.1250	50	44.85
Coke	570	0.1750	36	62.95
Concrete (breeze)	1440	0.6940	90	24.92
Concrete (cinder)	1762	0.5675	110	20.36
Concrete (gravel/ballast)	2240	0.4460	140	16.02
Concrete (mix)	2243	0.4458	140	16.00
Earth (dry)	1201	0.8326	75	29.88
Earth (moist)	1362	0.7342	85	26.34
Earth (mud, fluid)	1762	0.5675	110	20.36
Earth (top soil)	1600	0.6250	100	22.43
Earth (vegetable)	1230	0.8130	77	29.17
Flint	2590	0.3860	162	13.85
Flint pebbles	1682	0.5945	105	21.33
Granite (crushed, 32×25 mm)	1602	0.6242	100	22.40
Granite (lumps)	1538	0.6502	96	23.33
Granite (solid)	2563	0.3902	160	14.00
Gravel (coarse with sand)	1760	0.5680	110	20.39
Gravel (average)	1570	0.6369	98	22.85
Lime (briquettes)	961	1.0406	60	37.34
Lime (ground quicklime)	960	1.0420	60	37.38
Lime (hydrated, 200 mesh)	481	2.0790	30	74.60
Lime (quick, ground)	1041	0.9606	65	34.47
Lime (quick, lump)	1201	0.8326	75	29.88
Lime (slaked)	480	2.0830	30	74.75
Limestone	2643	0.3784	165	13.58
Limestone (lumps)	1522	0.6570	95	23.58
Loam	1600	0.6250	100	22.43
Marl	1760	0.5680	110	20.39
Media (filter)	880	1.1360	55	40.77
Mortar (wet)	2403	0.4161	150	14.93
Mud	1762	0.5675	110	20.36
Pebbles	1602	0.6242	100	22.40
Pitch	1160	0.8620	72	30.93

Densities of Various Materials

Building trade

Units	kg/m³	m³/tonne	lb/ft³	ft³/ton
Material	Metric density	Metric volume	Imperial density	Imperial volume
Pitch	1121	0.8921	70	32.01
Rubble (masonry)	2243	0.4458	140	16.00
Sand & gravel (dry)	1602	0.6242	100	22.40
Sand & gravel (wet)	1922	0.5203	120	18.67
Sand (clean pit, fine)	1440	0.6940	90	24.92
Sand (dry, loose)	1442	0.6935	90	24.88
Sand (foundry)	1522	0.6570	95	23.58
Sand (medium pit)	1530	0.6540	96	23.45
Sand (washed river)	1690	0.5920	106	21.23
Sand (wet, loose)	1762	0.5675	110	20.36
Sandstone (lumps)	1346	0.7429	84	26.66
Shale	2600	0.3850	162	13.80
Shale (crushed)	1442	0.6935	90	24.88
Slag	1510	0.6620	94	23.76
Slag (bank, crushed)	1281	0.7806	80	28.01
Slag (furnace, granulated)	1009	0.9911	63	35.56
Slate	2890	0.3460	180	12.42
Snow (compact)	520	1.9230	32	69.00
Snow (freshly fallen)	120	8.3330	7	299.01
Stone (basalt)	2770	0.3610	173	12.95
Stone (bath)	2000	0.5000	125	17.94
Stone (crushed)	1602	0.6242	100	22.40
Stone (granite)	2670	0.3750	167	13.44
Stone (Kentish rag)	2640	0.3790	165	13.59
Stone (limestone)	2410	0.4150	150	14.89
Stone (Portland)	2440	0.4100	152	14.71
Stone (Purbeck)	2600	0.3850	162	13.80
Stone (sandstone)	2330	0.4290	145	15.40
Stone (traprock)	2730	0.3660	170	13.14
Stone (whinstone)	2770	0.3610	173	12.95
Tar (liquid)	1145	0.8734	71	31.34
Trap rock (crushed)	1602	0.6242	100	22.40

Densities of Various Materials

Grain

Units	kg/m³	m³/tonne	lb/ft³	ft³/ton
Material	Metric density	Metric volume	Imperial density	Imperial volume
Barley	609	1.6420	38	58.92
Bran	256	3.9063	16	140.16
Brewer's grain (dry)	449	2.2272	28	79.91
Brewer's grain (wet)	881	1.1351	55	40.73
Corn (shelled)	721	1.3870	45	49.77
Flour (wheat)	641	1.5601	40	55.98
Oats	417	2.3981	26	86.05
Rice (hulled)	721	1.3870	45	49.77
Rice (rough)	577	1.7331	36	62.19
Rye	705	1.4184	44	50.90
Wheat	769	1.3004	48	46.66

Liquids

Units	kg/m³	m³/tonne	lb/ft³	ft³/ton
Material	Metric density	Metric volume	Imperial density	Imperial volume
Acid, nitric (91%)	1506	0.6640	94	23.83
Acid, sulphuric (87%)	1794	0.5574	112	20.00
Alcohol	785	1.2739	49	45.71
Benzene	737	1.3569	46	48.69
Gasoline	673	1.4859	42	53.32
Oils	929	1.0764	58	38.62
Paraffin	897	1.1148	56	40.00
Petrol	881	1.1351	55	40.73
Refined petrol	801	1.2484	50	44.80
Water (fresh)	1000	1.0000	62	35.88
Water (salt)	1025	0.9756	64	35.01

Miscellaneous

Units	kg/m³	m³/tonne	lb/ft³	ft³/ton
Material	Metric density	Metric volume	Imperial density	Imperial volume
Acid phosphate (pulv)	960	1.0417	60	37.38
Alfalfa (ground)	256	3.9063	16	140.16
Ammonium sulphate	850	1.1765	53	42.21
Apples	640	1.5625	40	56.06
Asbestos (shredded)	368	2.7174	23	97.50
Asbestos (solid)	2403	0.4161	150	14.93

Densities of Various Materials

Miscellaneous

Units	kg/m³	m³/tonne	lb/ft³	ft³/ton
Material	Metric density	Metric volume	Imperial density	Imperial volume
Asbestos (ground)	240	4.1667	15	149.51
Bagasse	1169	0.8554	73	30.69
Bakelite (moulded)	1362	0.7342	85	26.34
Bakelite (powdered)	561	1.7825	35	63.96
Baking powder	769	1.3004	48	46.66
Bark	240	4.1667	15	149.51
Barytes (crushed)	2883	0.3469	180	12.45
Barytes (powdered)	2162	0.4625	135	16.60
Bauxite (crushed)	1233	0.8110	77	29.10
Beans (castor)	577	1.7331	36	62.19
Beans (cocoa)	593	1.6863	37	60.51
Beans (navy)	817	1.2240	51	43.92
Beans (soya)	721	1.3870	45	49.77
Beets	721	1.3870	45	49.77
Bones	609	1.6420	38	58.92
Borax (fine)	881	1.1351	55	40.73
Calcium carbide (crushed)	1233	0.8110	77	29.10
Carbon (activated, fine, dry)	240	4.1667	15	149.51
Carbon (solid)	2162	0.4625	135	16.60
Carbon black (pellets)	416	2.4038	26	86.25
Charcoal	400	2.5000	25	89.70
Charcoal (ground)	160	6.2500	10	224.26
Chocolate (powder)	641	1.5601	40	55.98
Cinders (ashes, clinker)	641	1.5601	40	55.98
Cinders (blast furnace)	913	1.0953	57	39.30
Coal (anthracite)	961	1.0406	60	37.34
Coal (anthracite, solid)	1554	0.6435	97	23.09
Coal (bituminous)	801	1.2484	50	44.80
Coal (bituminous, solid)	1362	0.7342	85	26.34
Coconut (shredded)	368	2.7174	23	97.50
Coffee beans	481	2.0790	30	74.60
Coke (average lump)	481	2.0790	30	74.60
Cork (solid)	240	4.1667	15	149.51
Cork (ground)	144	6.9444	9	249.18
Cotton (lint)	96	10.4167	6	373.76
Cotton (seed & lint)	240	4.1667	15	149.51
Cottonseed (dry)	400	2.5000	25	89.70
Cottonseed (meal)	609	1.6420	38	58.92

Densities of Various Materials

Miscellaneous

Units	kg/m³	m³/tonne	lb/ft³	ft³/ton
Material	Metric density	Metric volume	Imperial density	Imperial volume
Dolomite (calcined)	641	1.5601	40	55.98
Dolomite (crushed)	1602	0.6242	100	22.40
Feldspar	1522	0.6570	95	23.58
Feldspar (ground)	1201	0.8326	75	29.88
Fibre (hard)	1394	0.7174	87	25.74
Fish (meal)	593	1.6863	37	60.51
Flaxseed	721	1.3870	45	49.77
Flour	561	1.7825	35	63.96
Fluorspar	1682	0.5945	105	21.33
Fluorspar (ground)	1314	0.7610	82	27.31
Foundry (loose sand)	1362	0.7342	85	26.34
Foundry (pressed sand)	1682	0.5945	105	21.33
Foundry (refuse)	1121	0.8921	70	32.01
Fuller's earth (dry)	561	1.7825	35	63.96
Fuller's earth (oily)	1041	0.9606	65	34.47
Glass	2563	0.3902	160	14.00
Glass (broken)	1442	0.6935	90	24.88
Glue	561	1.7825	35	63.96
Graphite (flakes)	640	1.5625	40	56.06
Greenstone (lumps)	1714	0.5834	107	20.93
Gypsum (crushed)	1522	0.6570	95	23.58
Gypsum (ground)	865	1.1561	54	41.48
Gypsum (lumps)	1281	0.7806	80	28.01
Gypsum (solid)	2243	0.4458	140	16.00
Hay (loose)	80	12.5000	5	448.52
Hay (packed)	384	2.6042	24	93.44
Hops (moist)	561	1.7825	35	63.96
Ice (solid)	913	1.0953	57	39.30
Ice (broken)	642	1.5576	40	55.89
Lead (red)	3684	0.2714	230	9.74
Lead (white)	4085	0.2448	255	8.78
Lead (white, dry)	1522	0.6570	95	23.58
Lignite (air-dried)	801	1.2484	50	44.80
Linseed (meal)	560	1.7857	35	64.07
Malt (wet)	1041	0.9606	65	34.47
Malt (whole, dry)	401	2.4938	25	89.48
Marble (crushed)	1522	0.6570	95	23.58
Marble (solid)	2723	0.3672	170	13.18

Densities of Various Materials

Miscellaneous

Units	kg/m³	m³/tonne	lb/ft³	ft³/ton
Material	Metric density	Metric volume	Imperial density	Imperial volume
Mica (ground)	1233	0.8110	77	29.10
Mica (solid)	2883	0.3469	180	12.45
Milk (powder)	320	3.1250	20	112.13
Oil (cake)	801	1.2484	50	44.80
Oyster shells	961	1.0406	60	37.34
Paper (writing/wrapping)	1281	0.7806	80	28.01
Peanuts (shelled)	593	1.6863	37	60.51
Peanuts (unshelled)	288	3.4722	18	124.59
Phosphate (rock lumps)	1522	0.6570	95	23.58
Phosphate (super)	1121	0.8921	70	32.01
Plaster of Paris	2243	0.4458	140	16.00
Potash (32 × 12 mm)	1121	0.8921	70	32.01
Potatoes (white)	769	1.3004	48	46.66
Pumice (ground)	673	1.4859	42	53.32
Quartz (lumps)	1362	0.7342	85	26.34
Refuse (average)	481	2.0790	30	74.60
Resin (vinyl)	481	2.0790	30	74.60
Rosin	1073	0.9320	67	33.44
Rubber (ground)	384	2.6042	24	93.44
Rubber (shredded scrap)	737	1.3569	46	48.69
Salt (coarse)	753	1.3280	47	47.65
Salt (fine)	1201	0.8326	75	29.88
Salt cake (coarse)	1442	0.6935	90	24.88
Sawdust (dry)	176	5.6818	11	203.87
Scale (rolling mill)	2002	0.4995	125	17.92
Sewage (raw sludge)	1041	0.9606	65	34.47
Sewage (screenings, drained)	881	1.1351	55	40.73
Shot (steel)	4005	0.2497	250	8.96
Silica (flour)	1041	0.9606	65	34.47
Slate (fine ground)	1362	0.7342	85	26.34
Slate (solid)	2723	0.3672	170	13.18
Snow (wet, old)	801	1.2484	50	44.80
Soap (chips)	160	6.2500	10	224.26
Soap (powder)	320	3.1250	20	112.13
Soda ash (dense)	1009	0.9911	63	35.56
Soda ash (light)	401	2.4938	25	89.48
Soda, bicarbonate	881	1.1351	55	40.73
Soyabean (flour)	433	2.3095	27	82.87

Densities of Various Materials

Miscellaneous

Units	kg/m³	m³/tonne	lb/ft³	ft³/ton
Material	Metric density	Metric volume	Imperial density	Imperial volume
Soyabean (meal)	641	1.5601	40	55.98
Soyabeans (cracked)	561	1.7825	35	63.96
Soyabeans (whole)	801	1.2484	50	44.80
Sugar (granulated)	849	1.1779	53	42.26
Sugar (powdered)	801	1.2484	50	44.80
Sugar cane	288	3.4722	18	124.59
Sugarbeet (dry pulp)	208	4.8077	13	172.51
Sugarbeet (wet pulp)	641	1.5601	40	55.98
Sulphur (lumps)	1281	0.7806	80	28.01
Sulphur (powdered)	881	1.1351	55	40.73
Talc (granulated)	961	1.0406	60	37.34
Talc (solid)	2723	0.3672	170	13.18
Tanbark (ground)	881	1.1351	55	40.73
Tobacco	320	3.1250	20	112.13
Tungsten carbide (powder)	4005	0.2497	250	8.96
Turf	401	2.4938	25	89.48
Wood chips	192	5.2083	12	186.88
Wood flour	368	2.7174	23	97.50
Zinc oxide (heavy)	529	1.8904	33	67.83
Zinc oxide (light)	240	4.1667	15	149.51

The periodic table

Period	I (1)	II (2)	3	4	5	6	7	8	9	10	11	12	III (13)	IV (14)	V (15)	VI (16)	VII (17)	VIII (18)
1	1 H 1.008																	2 He 4.003
2	3 Li 6.94	4 Be 9.01											5 B 10.81	6 C 12.01	7 N 14.01	8 O 16.00	9 F 19.00	10 Ne 20.18
3	11 Na 22.99	12 Mg 24.31											13 Al 26.98	14 Si 28.09	15 P 30.97	16 S 32.06	17 Cl 35.45	18 Ar 39.95
4	19 K 39.10	20 Ca 40.08	21 Sc 44.96	22 Ti 47.90	23 V 50.94	24 Cr 52.01	25 Mn 54.94	26 Fe 55.85	27 Co 58.93	28 Ni 58.71	29 Cu 63.54	30 Zn 65.37	31 Ga 69.72	32 Ge 72.59	33 As 74.92	34 Se 78.96	35 Br 79.91	36 Kr 83.80
5	37 Rb 85.47	38 Sr 87.62	39 Y 88.91	40 Zr 91.22	41 Nb 92.91	42 Mo 95.94	43 Tc 98.91	44 Ru 101.07	45 Rh 102.91	46 Pd 106.4	47 Ag 107.87	48 Cd 112.40	49 In 114.82	50 Sn 118.69	51 Sb 121.75	52 Te 127.60	53 I 126.90	54 Xe 131.30
6	55 Cs 132.91	56 Ba 137.34	71 Lu 174.97	72 Hf 178.49	73 Ta 180.95	74 W 183.85	75 Re 186.2	76 Os 190.2	77 Ir 192.2	78 Pt 195.09	79 Au 196.97	80 Hg 200.59	81 Tl 204.37	82 Pb 207.19	83 Bi 208.98	84 Po 210	85 At 210	86 Rn 222
7	87 Fr 223	88 Ra 226.03	103 Lr 257	104 Unq	105 Unp	106 Unh	107 Uns	108 Uno	109 Une									

Lanthanides	57 La 138.91	58 Ce 140.12	59 Pr 140.91	60 Nd 144.24	61 Pm 146.92	62 Sm 150.35	63 Eu 151.96	64 Gd 157.25	65 Tb 158.92	66 Dy 162.50	67 Ho 164.93	68 Er 167.26	69 Tm 168.93	70 Yb 173.04
Actinides	89 Ac 227.03	90 Th 232.04	91 Pa 231.04	92 U 238.03	93 Np 237.05	94 Pu 239.05	95 Am 241.06	96 Cm 247.07	97 Bk 249.08	98 Cf 251.08	99 Es 254.09	100 Fm 257.10	101 Md 258.10	102 No 255

Physical constants

APPROXIMATE VALUES OF SOME PHYSICAL CONSTANTS

Atoms and spectroscopy

Name	Constant	SI unit
Bohr magneton	$9.274\ 08 \times 10^{-24}$	joules/tesla
Bohr radius	$5.291\ 77 \times 10^{-11}$	metres
Charge/mass ratio of electron	$1.758\ 80 \times 10^{11}$	coulombs/kg
Classical radius of electron	$2.817\ 94 \times 10^{-15}$	metres
Compton wavelength of electron	$3.861\ 59 \times 10^{-13}$	metres
Gyromagnetic ratio of proton	$2.675\ 20 \times 10^{8}$	/second tesla
Nuclear magneton	$5.050\ 82 \times 10^{-27}$	joules/tesla
Planck's constant	$4.135\ 70 \times 10^{-15}$	eV seconds
Rydberg constant (fixed nucleus)	$1.097\ 37 \times 10^{7}$	/metre
Rydberg constant (Hydrogen)	$1.096\ 78 \times 10^{7}$	/metre
Thomson cross-section	$6.652\ 24 \times 10^{-29}$	metres
Zeeman effect	$4.668\ 60 \times 10^{1}$	/metre tesla

Energy, mass, and wavelength

Name	Constant	SI unit
Atomic mass unit (u)	$1.660\ 57 \times 10^{-27}$	kg
Electron rest mass	$0.511\ 00$	MeV
Electronvolt (eV)	$1.602\ 19 \times 10^{-19}$	joules
Electronvolt per molecule	$9.648\ 46 \times 10^{7}$	joules/kmole
Energy × wavelength	$1.239\ 85 \times 10^{-6}$	eV metres
Frequency/energy	$2.417\ 97 \times 10^{14}$	Hz/eV
Proton rest mass	$1.672\ 65 \times 10^{-27}$	kg
Quantum energy/frequency	$4.135\ 70 \times 10^{-15}$	eV/Hz
Quantum energy/wave number	$1.986\ 48 \times 10^{-25}$	joule metres
Wave number/energy	$8.065\ 48 \times 10^{5}$	/eV metre

Approximate Values of Some Physical Constants

Principal constants

Name	Constant	SI unit
Avogadro's number	$6.022\ 04 \times 10^{23}$	/mole
Electric constant	$8.854\ 19 \times 10^{-12}$	farads/metre
Electron mass in atomic mass units	$5.485\ 80 \times 10^{-4}$	units of (u)
Electron rest mass	$9.109\ 53 \times 10^{-31}$	kg
Electronic charge (e)	$1.602\ 19 \times 10^{-19}$	coulombs
Faraday constant (F)	$9.648\ 46 \times 10^{4}$	coulombs/mole
Gravitational acceleration (local, UK)	$9.806\ 65$	metres/sec^2
Planck's constant (h)	$6.626\ 18 \times 10^{-34}$	joule seconds
Proton rest mass	938.0	MeV
Speed of light in vacuum (c)	$2.997\ 92 \times 10^{8}$	metres/second
Standard temperature & pressure (stp)	$1.013\ 25 \times 10^{5}$	N/m^2 at 273.15 K
Universal constant of gravitation (G)	$6.672\ 0 \times 10^{-11}$	N metre^2/kg^2
Volume of 1 kmol of ideal gas (stp)	2.41	metres^3

Thermal constants

Name	Constant	SI unit
Boltzmann constant	$8.617\ 35 \times 10^{-5}$	eV/K
Boltzmann constant (k)	$1.380\ 66 \times 10^{-23}$	joules/K
Energy kT for T = 273.15 K	$0.023\ 54$	electron volt
Loschmidt constant	$2.686\ 75 \times 10^{25}$	/metre^3
Stefan–Boltzmann constant	$5.670\ 32 \times 10^{-8}$	W/(m^2 K^4)
Universal/Molar gas constant (R)	8.3143	kJ/(kmole K)

Mathematical formulae

INDICES

$$a^{p/q} = \sqrt[q]{a^p} = (a^p)^{1/q}, \quad a^{-m} = 1/a^m, \quad a^0 = 1$$

$$a^m a^n = a^{m+n}, \quad a^m/a^n = a^{m-n}, \quad (a^m)^n = a^{mn}.$$

LOGARITHMS

$$p = \log_a x \quad \text{if and only if} \quad x = a^p \ (a > 0)$$

$$e = \lim_{n \to \infty} \left(1 + \frac{1}{n}\right)^n = \sum_{i=0}^{\infty} \frac{1}{i!}$$

$$\ln x = \log_e x = \int_1^x \frac{dt}{t} \quad (x > 0)$$

$$\log_a x = \ln x / \ln a$$

$$\log_b x = \log_a x / \log_a b$$

$$\log_a x + \log_a y = \log_a xy, \quad \log_a x - \log_a y = \log_a(x/y)$$

$$\log_a x^n = n \log_a x$$

$$\lg x = \log_{10} x, \quad \text{lb } x = \log_2 x.$$

QUADRATIC EQUATIONS

The roots of the quadratic equation $ax^2 + bx + c = 0$ are given by

$$x = \frac{-b \pm \sqrt{(b^2 - 4ac)}}{2a}$$

The roots are real and distinct, real and equal, or complex according as $b^2 - 4ac$ is positive, zero, or negative respectively. The sum of the roots is $-b/a$; the product of the roots is c/a.

SERIES

TAYLOR'S THEOREM

$$f(a + x) = f(a) + f'(a)x + f''(a)\frac{x^2}{2!} + \cdots + f^{(n-1)}(a)\frac{x^{n-1}}{(n-1)!} + R_n$$

133

where R_n is the remainder which may be written in the forms

$$\text{(Lagrange)} \qquad R_n = \frac{f^{(n)}(a + \theta x)}{n!} x^n, \qquad 0 < \theta < 1$$

or

$$\text{(Cauchy)} \qquad R_n = \frac{f^{(n)}(a + \tilde{\theta}x)(1 - \tilde{\theta})^{n-1}}{(n-1)!} x^n, \qquad 0 < \tilde{\theta} < 1$$

When $a = 0$, Taylor's theorem gives Maclaurin's expansion

$$f(x) = f(0) + f'(0)x + f''(0)\frac{x^2}{2!} + f'''(0)\frac{x^3}{3!} + \cdots$$

BINOMIAL SERIES

$$(1 + x)^n = 1 + nx + \frac{n(n-1)}{2!}x^2 + \cdots + \frac{n(n-1)\cdots(n-r+1)}{r!}x^r + \cdots$$

The coefficient of x^r is denoted by

$$\binom{n}{r}, \quad {}^nC_r, \quad \text{or} \quad {}_nC_r.$$

The series terminates when n is a positive integer; otherwise it converges for $|x| < 1$ $(n < 0)$ and for $|x| \leqslant 1$ $(n \geqslant 0)$.

If n is a positive integer, then

$$\binom{n}{r} = \frac{n!}{r!(n-r)!} = \binom{n}{n-r}.$$

SUM OF ARITHMETIC PROGRESSION to n terms

$$a + (a + d) + (a + 2d) + \cdots + [a + (n-1)d] = \tfrac{1}{2}n[2a + (n-1)d]$$

SUM OF GEOMETRIC PROGRESSION to n terms

$$S_n = a + ar + ar^2 + \cdots + ar^{n-1} = \frac{a(1 - r^n)}{1 - r} = \frac{a(r^n - 1)}{r - 1}$$

Sum to infinity

$$\lim_{n \to \infty} S_n = a/(1 - r) \quad \text{for} \quad |r| < 1.$$

Series

SUMS OF POWERS OF THE NATURAL NUMBERS

$$\sum_{r=1}^{n} r = 1 + 2 + 3 + \cdots + n = \tfrac{1}{2}n(n+1)$$

$$\sum_{r=1}^{n} r^2 = 1^2 + 2^2 + 3^2 + \cdots + n^2 = \tfrac{1}{6}n(n+1)(2n+1)$$

$$\sum_{r=1}^{n} r^3 = 1^3 + 2^3 + 3^3 + \cdots + n^3 = \tfrac{1}{4}n^2(n+1)^2$$

COMMON POWER SERIES

$$e^x = 1 + x + \frac{x^2}{2!} + \frac{x^3}{3!} + \cdots \qquad \text{for all } x$$

$$\ln(1+x) = x - \frac{x^2}{2} + \frac{x^3}{3} - \frac{x^4}{4} + \cdots \qquad (-1 < x \leqslant 1)$$

$$\sin x = x - \frac{x^3}{3!} + \frac{x^5}{5!} - \frac{x^7}{7!} + \cdots \qquad \text{for all } x$$

$$\cos x = 1 - \frac{x^2}{2!} + \frac{x^4}{4!} - \frac{x^6}{6!} + \cdots \qquad \text{for all } x$$

$$\tan^{-1} x = x - \frac{x^3}{3} + \frac{x^5}{5} - \frac{x^7}{7} + \cdots \qquad (-1 \leqslant x \leqslant 1)$$

$$\tfrac{1}{2}\left(\frac{1+x}{1-x}\right) = x + \frac{x^3}{3} + \frac{x^5}{5} + \frac{x^7}{7} + \cdots \qquad (-1 < x < 1)$$

STIRLING'S FORMULA

$$n! = \sqrt{(2\pi n)}\left(\frac{n}{e}\right)^n\left(1 + \frac{1}{12n} + \frac{1}{288n^2} - \frac{139}{51840n^3} + \cdots\right)$$

TRIGONOMETRY

TRIANGLE

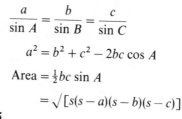

$$\frac{a}{\sin A} = \frac{b}{\sin B} = \frac{c}{\sin C}$$

$$a^2 = b^2 + c^2 - 2bc \cos A$$

$$\text{Area} = \tfrac{1}{2}bc \sin A$$

$$= \sqrt{[s(s-a)(s-b)(s-c)]}$$

where

$$s = \tfrac{1}{2}(a + b + c)$$

$$A + B + C = \pi = 180°$$

TRIGONOMETRIC IDENTITIES

$$\sin^2 A + \cos^2 A = 1, \quad \tan A = \frac{\sin A}{\cos A}$$

$$\tan^2 A + 1 = \sec^2 A$$

$$\sin(A \pm B) = \sin A \cos B \pm \cos A \sin B$$

$$\cos(A \pm B) = \cos A \cos B \mp \sin A \sin B$$

$$\tan(A \pm B) = \frac{\tan A \pm \tan B}{1 \mp \tan A \tan B}$$

$$2 \sin A \cos B = \sin(A + B) + \sin(A - B)$$

$$2 \cos A \sin B = \sin(A + B) - \sin(A - B)$$

$$2 \cos A \cos B = \cos(A + B) + \cos(A - B)$$

$$2 \sin A \sin B = \cos(A - B) - \cos(A + B)$$

$$\sin A + \sin B = 2 \sin \tfrac{1}{2}(A + B) \cos \tfrac{1}{2}(A - B)$$

$$\sin A - \sin B = 2 \cos \tfrac{1}{2}(A + B) \sin \tfrac{1}{2}(A - B)$$

$$\cos A + \cos B = 2 \cos \tfrac{1}{2}(A + B) \cos \tfrac{1}{2}(A - B)$$

$$\cos A - \cos B = -2 \sin \tfrac{1}{2}(A + B) \sin \tfrac{1}{2}(A - B)$$

$$\sin^2 A = \tfrac{1}{2}(1 - \cos 2A); \quad \cos^2 A = \tfrac{1}{2}(1 + \cos 2A)$$

$$\sin \theta = \frac{2t}{1 + t^2}, \quad \cos \theta = \frac{1 - t^2}{1 + t^2}, \quad \text{where} \quad t = \tan \tfrac{1}{2}\theta$$

$$\lim_{\theta \to 0} \frac{\sin \theta}{\theta} = 1$$

$$\sin 2\theta = 2 \sin \theta \cos \theta$$

$$\cos 2\theta = \cos^2 \theta - \sin^2 \theta = 1 - 2 \sin^2 \theta = 2 \cos^2 \theta - 1$$

$$\tan 2\theta = \frac{2 \tan \theta}{1 - \tan^2 \theta}$$

HYPERBOLIC FUNCTIONS

$$\cosh x = \tfrac{1}{2}(e^x + e^{-x}), \qquad \sinh x = \tfrac{1}{2}(e^x - e^{-x})$$

$$\tanh x = \frac{e^x - e^{-x}}{e^x + e^{-x}} = \frac{1 - e^{-2x}}{1 + e^{-2x}}$$

$$\coth x = \frac{e^x + e^{-x}}{e^x - e^{-x}} = \frac{1 + e^{-2x}}{1 - e^{-2x}}$$

$$\operatorname{sech} x = \frac{2}{e^x + e^{-x}}, \qquad \operatorname{cosech} x = \frac{2}{e^x - e^{-x}}$$

INVERSE HYPERBOLIC FUNCTIONS

$$\cosh^{-1} x = \ln\left[x + \sqrt{(x^2 - 1)}\right] \quad (|x| \geqslant 1)$$

$$\sinh^{-1} x = \ln\left[x + \sqrt{(x^2 + 1)}\right]$$

$$\tanh^{-1} x = \tfrac{1}{2}\ln\frac{1 + x}{1 - x} \quad (|x| < 1)$$

COORDINATE GEOMETRY

Distance between two points (x_1, y_1) and (x_2, y_2)

$$d = \sqrt{[(x_1 - x_2)^2 + (y_1 - y_2)^2]}$$

Equation of straight line of gradient m with intercept c on y axis

$$y = mx + c$$

Gradient of line between two points (x_1, y_1) and (x_2, y_2)

$$m = \frac{y_2 - y_1}{x_2 - x_1}$$

Equation of straight line of gradient m through (x_1, y_1)

$$y - mx = y_1 - mx_1$$

Coordinate Geometry

CONIC SECTIONS (see Table 1)

Table 1

	Circle	Ellipse
Cartesian equation	$x^2 + y^2 = a^2$	$\dfrac{x^2}{a^2} + \dfrac{y^2}{b^2} = 1$
Eccentricity ε	0	$\sqrt{\left(1 - \dfrac{b^2}{a^2}\right)}$
Focal distance OF	0	$a\varepsilon$
	Hyperbola	**Parabola**
Cartesian equation	$\dfrac{x^2}{a^2} - \dfrac{y^2}{b^2} = 1$	$y^2 = 4ax$
Eccentricity ε	$\sqrt{\left(1 + \dfrac{b^2}{a^2}\right)}$	1
Focal distance OF	$a\varepsilon$	a

General forms for the equation of a circle

$$(x - h)^2 + (y - k)^2 = r^2, \quad \text{centre } (h, k) \text{ and radius } r$$

$$x^2 + y^2 + 2gx + 2fy + c = 0, \quad \text{centre } (-g, -f) \text{ and}$$
$$\text{radius } \sqrt{(g^2 + f^2 - c)}$$

Coordinate Geometry

Form for the equation of a rectangular hyperbola with the cartesian axes as asymptotes

$$xy = c^2$$

General equation of parabola with axis parallel to the y axis

$$y = ax^2 + bx + c$$

CALCULUS

DIFFERENTIATION: NOTATION

Let f be a real function, with x and y real variables related by $y = f(x)$. The derivative function is denoted by f' or Df. The value of the derivative at x is denoted by

$$\frac{dy}{dx}, \; f'(x), \; y', \; Df(x), \; \text{or} \; Dy$$

DERIVATIVES

Table 2. In the table, u and v are functions of x, and a is a constant.

$f(x)$	$f'(x)$	$f(x)$	$f'(x)$
a	0	$\sin x$	$\cos x$
au	au'	$\cos x$	$-\sin x$
$u + v$	$u' + v'$	$\tan x$	$\sec^2 x$
uv	$vu' + uv'$	$\operatorname{cosec} x$	$-\cot x \operatorname{cosec} x$
$\dfrac{u}{v}$	$\dfrac{vu' - uv'}{v^2}$	$\sec x$	$\tan x \sec x$
		$\cot x$	$-\operatorname{cosec}^2 x$
$g(u)$	$g'(u)u'$	$\sinh x$	$\cosh x$
u^n	$nu^{n-1}u'$	$\cosh x$	$\sinh x$
		$\tanh x$	$\operatorname{sech}^2 x$
$\ln u$	$\dfrac{u'}{u}$	$\operatorname{cosech} x$	$-\coth x \operatorname{cosech} x$
		$\operatorname{sech} x$	$-\tanh x \operatorname{sech} x$
x^n	nx^{n-1}	$\coth x$	$-\operatorname{cosech}^2 x$
e^x	e^x	$\sin^{-1}(x/a)$	$(a^2 - x^2)^{-1/2}$
		$\cos^{-1}(x/a)$	$-(a^2 - x^2)^{-1/2}$
$\ln x$	$\dfrac{1}{x}$	$\tan^{-1}(x/a)$	$a/(a^2 + x^2)$

Calculus

INDEFINITE INTEGRALS

Table 3. The constant of integration is omitted in each case; u and v are functions of x, and a is a constant.

$f(x)$	$\int f(x)\,dx$
au	$a \int u\,dx$
$u + v$	$\int u\,dx + \int v\,dx$
uv	$u \int v\,dx - \int \dfrac{du}{dx}\left(\int v\,dx\right)dx$
$g(u)\dfrac{du}{dx}$	$\int g(u)\,du$
$g(ax)$	$\dfrac{1}{a}\int g(u)\,du \quad (u = ax)$
$g(x + a)$	$\int g(u)\,du \quad (u = x + a)$
x^n	$\dfrac{1}{n+1}x^{n+1} \quad (n \neq -1)$
x^{-1}	$\ln x$
e^x	e^x
$\ln x$	$x(\ln x - 1)$
$\sin x$	$-\cos x$
$\cos x$	$\sin x$
$\tan x$	$-\ln \cos x$
$\operatorname{cosec} x$	$\ln \tan (x/2)$
$\sec x$	$\ln (\sec x + \tan x)$
$\cot x$	$\ln \sin x$
$\sin^2 x$	$\frac{1}{2}(x - \frac{1}{2}\sin 2x)$
$\cos^2 x$	$\frac{1}{2}(x + \frac{1}{2}\sin 2x)$
$\tan^2 x$	$\tan x - x$
$\operatorname{cosec}^2 x$	$-\cot x$
$\sec^2 x$	$\tan x$
$\cot^2 x$	$-x - \cot x$
$x \sin x$	$\sin x - x \cos x$
$x \cos x$	$\cos x + x \sin x$
$\sin^{-1} x$	$x \sin^{-1} x + (1 - x^2)^{1/2}$
$\cos^{-1} x$	$x \cos^{-1} x - (1 - x^2)^{1/2}$
$\tan^{-1} x$	$x \tan^{-1} x - \frac{1}{2}\ln (1 + x^2)$
$\sinh x$	$\cosh x$

140

Calculus

$\cosh x$	$\sinh x$
$\tanh x$	$\ln \cosh x$
$\operatorname{cosech} x$	$\ln \tanh \frac{1}{2}x$
$\operatorname{sech} x$	$\tan^{-1}(\sinh x)$
$\coth x$	$\ln \sinh x$
$\sinh^2 x$	$\frac{1}{2}(-x + \frac{1}{2}\sinh 2x)$
$\cosh^2 x$	$\frac{1}{2}(x + \frac{1}{2}\sinh 2x)$
$\tanh^2 x$	$x - \tanh x$
$\operatorname{cosech}^2 x$	$-\coth x$
$\operatorname{sech}^2 x$	$\tanh x$
$\coth^2 x$	$x - \coth x$
$x \sinh x$	$x \cosh x - \sinh x$
$x \cosh x$	$x \sinh x - \cosh x$
$\sinh^{-1} x$	$x \sinh^{-1} x - (x^2 + 1)^{1/2}$
$\cosh^{-1} x$	$x \cosh^{-1} x - (x^2 - 1)^{1/2}$
$\tanh^{-1} x$	$x \tanh^{-1} x + \frac{1}{2}\ln(1 - x^2)$
$(x^2 + a^2)^{-1}$	$\dfrac{1}{a}\tan^{-1}(x/a)$
$(x^2 - a^2)^{-1}$	$\dfrac{1}{2a}\ln\dfrac{x-a}{x+a}$
$(a^2 - x^2)^{-1}$	$\dfrac{1}{2a}\ln\dfrac{a+x}{a-x}$
$(x^2 + a^2)^{1/2}$	$\frac{1}{2}[x(x^2 + a^2)^{1/2} + a^2 \sinh^{-1}(x/a)]$
$(x^2 - a^2)^{1/2}$	$\frac{1}{2}[x(x^2 - a^2)^{1/2} - a^2 \cosh^{-1}(x/a)]$
$(a^2 - x^2)^{1/2}$	$\frac{1}{2}[x(a^2 - x^2)^{1/2} + a^2 \sin^{-1}(x/a)]$
$(x^2 + a^2)^{-1/2}$	$\sinh^{-1}(x/a)$
$(x^2 - a^2)^{-1/2}$	$\cosh^{-1}(x/a)$
$(a^2 - x^2)^{-1/2}$	$\sin^{-1}(x/a)$
$a/x(x^2 - a^2)^{1/2}$	$\sec^{-1}(x/a)$

DEFINITE INTEGRALS

Legendre's normal *elliptic integrals* include:

$$F(\vartheta, k) = \int_0^\vartheta \frac{\mathrm{d}\theta}{(1 - k^2 \sin^2 \theta)^{\frac{1}{2}}} \quad \text{(first kind)}$$

$$E(\vartheta, k) = \int_0^\vartheta (1 - k^2 \sin^2 \theta)^{\frac{1}{2}} \, \mathrm{d}\theta \quad \text{(second kind)}$$

where $0 < k < 1$. The 'complete' forms of these are

$$F(\tfrac{1}{2}\pi, k) = F(k) = \int_0^{\frac{1}{2}\pi} \frac{\mathrm{d}\theta}{(1 - k^2 \sin^2 \theta)^{\frac{1}{2}}}$$

$$= \frac{\pi}{2}\left[1 + \frac{1^2}{2^2}k^2 + \left(\frac{1.3}{2.4}\right)^2 k^4 + \cdots\right] \quad (k^2 < 1)$$

$$E(\tfrac{1}{2}\pi, k) = E(k) = \int_0^{\frac{1}{2}\pi} (1 - k^2 \sin^2 \theta)^{\frac{1}{2}} \, d\theta$$

$$= \frac{\pi}{2}\left[1 - \frac{1^2}{2^2}\frac{k^2}{1} - \left(\frac{1.3}{2.4}\right)^2 \frac{k^4}{3} - \cdots\right] \quad (k^2 < 1)$$

The *error function* erf is given by

$$\operatorname{erf} x = \frac{2}{\sqrt{\pi}} \int_0^x e^{-u^2} \, du = \frac{2}{\sqrt{\pi}}\left(x - \frac{x^3}{3} + \frac{1}{2!}\frac{x^5}{5} - \frac{1}{3!}\frac{x^7}{7} + \cdots\right)$$

The *gamma function* $\Gamma(\cdot)$ is given by

$$\Gamma(x) = \int_0^\infty u^{x-1} e^{-u} \, du = \int_0^1 \left(\ln \frac{1}{u}\right)^{x-1} du \quad (x > 0)$$

$$\Gamma(n+1) = n! \quad (n \text{ an integer}), \quad \Gamma(\tfrac{1}{2}) = \sqrt{\pi}$$

$$\int_0^\infty e^{-x} \ln \frac{1}{x} \, dx = \gamma \quad (\approx 0.57721, \text{ Euler's constant})$$

$$\int_0^\infty \frac{x^{m-1}}{1 + x^n} \, dx = \frac{\pi}{n}\Big/\sin \frac{m\pi}{n} \quad (0 < m < n)$$

$$\int_0^\infty e^{-ax} \, dx = \frac{1}{a} \quad (a > 0)$$

$$\int_0^\infty x^n e^{-ax} \, dx = \frac{\Gamma(n+1)}{a^{n+1}} \quad (n > -1, a > 0)$$

$$\int_0^\infty x e^{-x^2} \, dx = \tfrac{1}{2}$$

$$\int_0^\infty x^2 e^{-x^2} \, dx = \tfrac{1}{4}\sqrt{\pi}$$

$$\int_0^{\frac{1}{2}\pi} \sin^n x \, dx = \int_0^{\frac{1}{2}\pi} \cos^n x \, dx = \frac{\sqrt{\pi}\,\Gamma(\tfrac{1}{2}n + \tfrac{1}{2})}{2\Gamma(\tfrac{1}{2}n + 1)} \quad (n > -1)$$

Calculus

$$\int_0^\infty \frac{\sin ax}{x}\,dx = \begin{cases} \frac{1}{2}\pi & (a > 0) \\ 0 & (a = 0) \\ -\frac{1}{2}\pi & (a < 0) \end{cases}$$

$$\int_0^\infty e^{-ax} \sin bx\,dx = \frac{b}{a^2 + b^2} \quad (a > 0)$$

$$\int_0^\infty e^{-ax} \cos bx\,dx = \frac{a}{a^2 + b^2} \quad (a > 0)$$

When a and b are integers,

$$\int_0^\pi \sin ax \cos bx\,dx = \begin{cases} 0 & (a - b\text{ even}) \\ 2a/(a^2 - b^2) & (a - b\text{ odd}) \end{cases}$$

$$\int_0^\pi \sin ax \sin bx\,dx = \int_0^\pi \cos ax \cos bx\,dx = \begin{cases} 0 & (a \neq b) \\ \frac{1}{2}\pi & (a = b) \end{cases}$$

$$\int_0^{\frac{1}{2}\pi} \sin^m x \cos^n x\,dx = \frac{\Gamma(\frac{1}{2}m + \frac{1}{2})\Gamma(\frac{1}{2}n + \frac{1}{2})}{2\Gamma(\frac{1}{2}m + \frac{1}{2}n + 1)}$$

NUMERICAL ANALYSIS

SOLUTION OF ALGEBRAIC EQUATION $f(x) = 0$, with x_n the nth estimate

$$x_{n+1} = x_n - f(x_n)/f'(x_n) \qquad \text{(Newton's method)}$$

$$x_{n+1} = \frac{x_{n-1}f(x_n) - x_n f(x_{n-1})}{f(x_n) - f(x_{n-1})} \quad \text{(secant method)}$$

NUMERICAL INTEGRATION (Simpson's rule)

$$\int_a^{a+nh} f(x)\,dx \approx$$

$$\frac{h}{3}\left(f(a) + f(a + nh) + 2\sum_{i=1}^{\frac{1}{2}n-1} f(a + 2ih) + 4\sum_{i=1}^{\frac{1}{2}n} f(a + (2i - 1)h) \right)$$

where n is an even positive integer

MENSURATION

LENGTH OF A PLANE CURVE
(cartesian form: $y = f(x)$ between $x = a$ and $x = b$)

$$s = \int_a^b \left[1 + \left(\frac{dy}{dx} \right)^2 \right]^{\frac{1}{2}} dx$$

(polar form: $r = f(\theta)$ between $\theta = \alpha$ and $\theta = \beta$)

$$s = \int_\alpha^\beta \left[r^2 + \left(\frac{dr}{d\theta} \right)^2 \right]^{\frac{1}{2}} d\theta$$

(parametric cartesian form: $(x, y) = (f(t), g(t))$ between $t = p$ and $t = q$)

$$s = \int_p^q \left[\left(\frac{dx}{dt} \right)^2 + \left(\frac{dy}{dt} \right)^2 \right]^{\frac{1}{2}} dt$$

PLANE AREA
The area bounded by the cartesian curve $y = f(x)$, the ordinates $x = a$ and $x = b$, and the x axis ($y = 0$) is

$$\int_a^b y \, dx$$

The area bounded by the polar curve $r = f(\theta)$, and the radial lines $\theta = \alpha$ and $\theta = \beta$ is

$$\tfrac{1}{2} \int_\alpha^\beta r^2 \, d\theta$$

SURFACE OF REVOLUTION
The area of the surface generated by one complete revolution, about the x axis, of the curve $y = f(x)$ from $x = a$ to $x = b$ is

$$2\pi \int_a^b y \left[1 + \left(\frac{dy}{dx} \right)^2 \right]^{\frac{1}{2}} dx$$

RADIUS OF CURVATURE
(for a plane curve at a point (x, y))

$$\rho = \left[1 + \left(\frac{dy}{dx} \right)^2 \right]^{\frac{1}{2}} \bigg/ \frac{d^2 y}{dx^2}$$

144

Mensuration

VOLUME OF REVOLUTION

The volume swept out by one complete revolution, about the x axis, of the area bounded by the curve $y = f(x)$, the ordinates $x = a$ and $x = b$, and the x axis is

$$\pi \int_a^b y^2 \, dx$$

MEANS AND MOMENTS

The mean value and root-mean-square value (RMS) of y, where $y = f(x)$, between $x = a$ and $x = b$, are respectively

$$\frac{1}{b-a} \int_a^b y \, dx, \qquad \left(\frac{1}{b-a} \int_a^b y^2 \, dx \right)^{\frac{1}{2}}$$

The coordinates $(\bar{x}, \bar{y})$ of the centroid G of the region of area A lying within the curve $y = f(x)$, the ordinates $x = a$ and $x = b$, and the x axis are

$$\bar{x} = \frac{1}{A} \int_a^b xy \, dx, \qquad \bar{y} = \frac{1}{A} \int_a^b \tfrac{1}{2} y^2 \, dx$$

and, for the same region, the radii of gyration, k_{GX} and k_{GY}, about axes GX and GY through G parallel to the x and y axes, are given by

$$k_{GX}^2 = \frac{1}{A} \int_a^b \tfrac{1}{3} y^3 \, dx - \bar{y}^2, \qquad k_{GY}^2 = \frac{1}{A} \int_a^b x^2 y \, dx - \bar{x}^2$$

PARTICULAR PLANE FIGURES AND BODIES IN THREE DIMENSIONS

See Tables 4 and 5. In the tables, $\bar{x}$, $\bar{y}$, and $\bar{z}$ are the coordinates of the centroid G with respect to the x, y, and z axes, and the radii of gyration k_{GX}, k_{GY}, and k_{GZ} are about axes through G parallel to the x, y, and z axes respectively, for a uniform lamina or body.

PERPENDICULAR-AXIS THEOREM FOR A LAMINA

For a lamina with radii of gyration k_{OX} and k_{OY} about perpendicular axes OX and OY in the plane of the lamina, the radius of gyration k_{OZ} about an axis through O perpendicular to the plane of the lamina is given by

$$k_{OZ}^2 = k_{OX}^2 + k_{OY}^2$$

145

Table 4

Figure		Area	$\bar{x}$	$\bar{y}$	k_{Gx}^2	k_{Gy}^2
Rectangle		bh	$\frac{1}{2}b$	$\frac{1}{2}h$	$\frac{1}{12}h^2$	$\frac{1}{12}b^2$
Triangle		$\frac{1}{2}(a+b)h$	$\frac{1}{3}b - \frac{1}{3}a$	$\frac{1}{3}h$	$\frac{1}{18}h^2$	$\frac{1}{18}(a^2 + ab + b^2)$
Circle of radius a, centred at O		πa^2	0	0	$\frac{1}{4}a^2$	$\frac{1}{4}a^2$
Semicircle		$\frac{1}{2}\pi a^2$	0	$4a/3\pi$	$\frac{1}{4}a^2 - \bar{y}^2$	$\frac{1}{4}a^2$
Sector of circle		$\frac{1}{2}a^2\alpha$	$\dfrac{4a\sin\frac{1}{2}\alpha}{3\alpha}$	0	$\frac{1}{4}a^2\left(1 - \dfrac{\sin 2\alpha}{2\alpha}\right)$	$\frac{1}{4}a^2\left(1 + \dfrac{\sin\alpha}{\alpha}\right) - \bar{x}^2$
Ellipse		πab	0	0	$\frac{1}{4}b^2$	$\frac{1}{4}a^2$
Semiellipse		$\frac{1}{2}\pi ab$	0	$4b/3\pi$	$\frac{1}{4}b^2 - \bar{y}^2$	$\frac{1}{4}a^2$
Parabolic segment		$\frac{4}{3}ah$	0	$\frac{2}{5}h$	$\frac{3}{7}\bar{y}^2$	$\frac{1}{5}a^2$
Thin rod		length $2a$	0	0	0	$\frac{1}{3}a^2$
Thin hoop of radius a, centred at O		length $2\pi a$	0	0	$\frac{1}{2}a^2$	$\frac{1}{2}a^2$

Body	Area*	Volume	$\bar{x}$	$\bar{y}$	$\bar{z}$	k^2_{Gx}	k^2_{Gx}	k^2_{Gz}
Rectangular prism	—	abc	$\frac{1}{2}a$	$\frac{1}{2}b$	$\frac{1}{2}c$	$\frac{1}{12}(b^2+c^2)$	$\frac{1}{12}(c^2+a^2)$	$\frac{1}{12}(a^2+b^2)$
Sphere of radius a, centred at O	$4\pi a^2$	$\frac{4}{3}\pi a^3$	0	0	0	$\frac{2}{5}a^2$	k^2_{Gx}	k^2_{Gx}
Hollow sphere	$4\pi a^2$	$\frac{4}{3}\pi(a^3-b^3)$	0	0	0	$\dfrac{2(a^5-b^5)}{5(a^3-b^3)}$	k^2_{Gx}	k^2_{Gx}
Thin spherical shell of radius a, centred at O	$4\pi a^2$	0	0	0	0	$\frac{2}{3}a^2$	k^2_{Gx}	k^2_{Gx}
Hemisphere	$2\pi a^2$	$\frac{2}{3}\pi a^3$	0	0	$\frac{3}{8}a$	$\dfrac{83a^2}{320}$	k^2_{Gx}	$\frac{2}{5}a^2$
Right circular cylinder	$2\pi ah$	$\pi a^2 h$	0	0	$\frac{1}{2}h$	$\frac{1}{4}a^2+\frac{1}{12}h^2$	k^2_{Gx}	$\frac{1}{2}a^2$
Right circular cone	$\pi a(a^2+h^2)^{\frac{1}{2}}$	$\frac{1}{3}\pi a^2 h$	0	0	$\frac{1}{4}h$	$\dfrac{3(4a^2+h^2)}{80}$	k^2_{Gx}	$\frac{3}{10}a^2$
Ellipsoid bounded by the surface $\dfrac{x^2}{a^2}+\dfrac{y^2}{b^2}+\dfrac{z^2}{c^2}=1$	**	$\frac{4}{3}\pi abc$	0	0	0	$\frac{1}{5}(b^2+c^2)$	$\frac{1}{5}(c^2+a^2)$	$\frac{1}{5}(a^2+b^2)$
Segment of thin spherical shell of radius r	$2\pi rh$	0	0	0	$\frac{1}{2}h$	$\frac{1}{2}rh-\frac{1}{12}h^2$	k^2_{Gx}	$rh-\frac{1}{3}h^2$

* Area refers to external curved surface only. ** Area not expressible in terms of elementary functions.

Means and Moments

PARALLEL-AXIS THEOREM FOR A BODY

For a body with radius of gyration k_{GX} about an axis through G, the radius of gyration $k_{OX'}$ about an axis OX', parallel to GX and distant a from it, is given by

$$k_{OX'}^2 = k_{GX}^2 + a^2$$

MOMENTS OF AREA, VOLUME, AND MASS

About any particular axis, the first moment of area (resp. volume, mass) is the product of the area (resp. volume, mass) with the distance of its centroid from the given axis, and the second moment is the product of the area (resp. volume, mass) with the squared radius of gyration about the axis. The moment of inertia I is the second moment of mass, i.e.

$$I = mk^2$$

where m is the mass and k is the radius of gyration.

STATISTICS

MEAN AND VARIANCE

The average (also known as the mean) of n values x_i of a random variable x is

$$\bar{x} = \frac{1}{n} \sum_{i=1}^{n} x_i$$

which is an unbiased estimate of the population mean, and the variance is

$$s^2 = \frac{1}{n} \sum_{i=1}^{n} (x_i - \bar{x})^2 = \frac{1}{n} \sum_{i=1}^{n} x_i^2 - \bar{x}^2$$

which is a maximum-likelihood estimate of the population variance; for a large population, an unbiased estimate, preferred in most cases, is

$$s_u^2 = \frac{1}{n-1} \sum_{i=1}^{n} (x_i - \bar{x})^2 = \frac{n}{n-1} s^2$$

If a random variable has probability density function f, then its expectation (also known as mean) and variance are given respectively by

$$\mu = \int_{-\infty}^{\infty} tf(t)\,dt, \qquad \sigma^2 = \int_{-\infty}^{\infty} (t-\mu)^2 f(t)\,dt$$

The standard deviation is the square root of the variance.

If the origin and scaling of data is changed so that a new random variable x' is defined by

$$x' = ax + b$$

then

$$\mu' = a\mu + b, \quad \sigma'^2 = a^2\sigma^2$$

where the primes indicate the mean and variance of x'.

For a large population with variance σ^2, the mean of a random sample of size n is a random variable with variance σ^2/n.

For discrete distributions, with p_k representing the probability that the random variable x takes the value x_k, the mean and variance are respectively

$$\mu = \sum_k p_k x_k, \qquad \sigma^2 = \sum_k p_k(x_k - \mu)^2$$

DISTRIBUTIONS

Discrete distributions

The random variable x is distributed binomially (*binomial* distribution) with parameters n and p if

$$p_k = P(x = k) = \begin{cases} \binom{n}{k} p^k(1-p)^{n-k} & (k = 0, \ldots, n) \\ 0 & \text{otherwise} \end{cases}$$

where n is an integer ($\geqslant 0$) and $0 \leqslant p \leqslant 1$. For the binomial distribution, the mean and variance are respectively

$$\mu = np, \qquad \sigma^2 = np(1-p)$$

The binomial probability p_k is the probability that an event having probability p of occurring in a single trial will occur just k times in n independent trials.

Statistics

Binomial distribution (n=5, p=2/5)

Poisson distribution (μ=2.5)

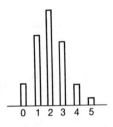

Rectangular distribution

Exponential distribution

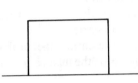

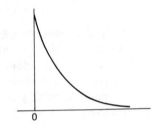

Normal distribution

Lognormal distribution

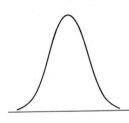

Student distribution

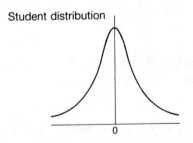

Statistics

The *Poisson* distribution with parameter μ (> 0) is given by the probabilities

$$p_k = \mathrm{P}(x = k) = \begin{cases} \mathrm{e}^{-\mu} \mu^k / k! & (k = 0, 1, \ldots) \\ 0 & \text{otherwise} \end{cases}$$

For this distribution, the mean and variance are both equal to μ. The Poisson is the limit of the binomial, as $n \to \infty$ with $p = \mu/n$, and typically represents the frequency of relatively rare events in large samples.

Continuous distributions

The *rectangular* distribution with parameters a and b ($> a$) has probability density function f given by

$$f(t) = \begin{cases} 1/(b - a) & \text{if } a \leqslant t \leqslant b \\ 0 & \text{otherwise} \end{cases}$$

The mean and variance for this distribution are respectively

$$\mu = \tfrac{1}{2}(a + b), \qquad \sigma^2 = \tfrac{1}{12}(b - a)^2$$

A rectangularly distributed quantity often arises as the residue after the sectioning of a larger random quantity, e.g. the amount of a product in a container on a filling machine at any instant.

The *exponential* distribution with parameter μ, which has mean μ and variance μ^2, has probability density

$$f(t) = \begin{cases} \mu^{-1} \mathrm{e}^{-t/\mu} & \text{if } t \geqslant 0 \\ 0 & \text{if } t < 0 \end{cases}$$

It is often used to model the time up to a random event, e.g. a survival time or a time between arrivals.

The normal (or Gauss) distribution with mean μ and variance σ^2 has probability density

$$f(t) = \frac{1}{(2\pi)^{\frac{1}{2}} \sigma} \exp \frac{-(t - \mu)^2}{2\sigma^2} \quad (-\infty < t < \infty)$$

It is used to represent variations that arise as the combination of a large number of independent small variations, and applies to a very wide range of natural phenomena and complex synthetic phenomena. The standard normal distribution is normal with mean 0 and variance 1.

A distribution related to the normal is the lognormal: a random variable is distributed lognormally if its natural logarithm is distributed normally—say, with mean μ and variance σ^2. In this case, the lognormal mean and variance are respectively

$$e^{\mu + \frac{1}{2}\sigma^2}, \qquad e^{2\mu + \sigma^2}(e^{\sigma^2} - 1)$$

This distribution is often appropriate in modelling the variation of quantities that are necessarily positive but which occur over a very wide range of relative sizes, e.g. sizes of geological bodies.

The Student (or 't') distribution with parameter v (> 0) has mean zero and variance $v/(v - 2)$ for $v > 2$ (the variance is infinite for $v \leqslant 2$, and the mean undefined for $v \leqslant 1$). Its probability density is

$$f(t) = \frac{\Gamma(\frac{1}{2}v + \frac{1}{2})}{(\pi v)^{\frac{1}{2}}\Gamma(\frac{1}{2}v)} \bigg/ \left(1 + \frac{t^2}{v}\right)^{\frac{1}{2}v + \frac{1}{2}} \qquad (-\infty < t < \infty)$$

It is used to compute confidence intervals for an estimate of the population mean calculated from sample data, where the population has unknown mean and variance but is assumed to be normal. The statistic

$$(m - \mu)n^{\frac{1}{2}}/s_u$$

has this distribution, where m is the mean of a sample of size n from a normal distribution of mean μ, and s_u^2 is the unbiased estimate of the variance from the sample; here $v = n - 1$. For large values of v, the Student approximates the standard normal.

Selected engineering formulae

THERMODYNAMICS AND FLUID MECHANICS

NOTATION

c_p specific heat capacity at constant pressure (J kg^{-1} K^{-1})
c_v specific heat capacity at constant volume (J kg^{-1} K^{-1})
γ ratio of specific heats c_p/c_v
g specific free energy (J kg^{-1})

Thermodynamics and Fluid Mechanics

h specific enthalpy (J kg^{-1})
κ compressibility (m^2 N^{-1})
m mass (kg)
η dynamic viscosity (N s m^{-2})
v kinematic viscosity (m^2 s^{-1})
p absolute pressure (N m^{-2})
q specific heat input (J kg^{-1})
R gas constant (J kg^{-1} K^{-1})
ρ mass density (kg m^{-3})
s specific entropy (J kg^{-1} K^{-1})
T absolute temperature (K)
u specific internal energy (J kg^{-1})
v specific volume (m^3 kg^{-1})
V volume (m^3)
w specific work output (J kg^{-1})

THERMODYNAMIC RELATIONS

Basic relations

Enthalpy $h = u + pv$
Helmholtz function $f = u - Ts$
Gibbs function $g = h - Ts$

$$dh = t \ ds + v \ dp, \qquad df = -p \ dv - s \ dT, \qquad dg = v \ dp - s \ dT$$

Maxwell's relations

$$\left(\frac{\partial T}{\partial v}\right)_s = -\left(\frac{\partial p}{\partial s}\right)_v, \qquad \left(\frac{\partial T}{\partial p}\right)_s = \left(\frac{\partial v}{\partial s}\right)_p$$

$$\left(\frac{\partial p}{\partial T}\right)_v = \left(\frac{\partial s}{\partial v}\right)_T, \qquad \left(\frac{\partial v}{\partial T}\right)_p = -\left(\frac{\partial s}{\partial p}\right)_T$$

Specific heats

$$c_p = \left(\frac{\partial h}{\partial T}\right)_p, \qquad c_v = \left(\frac{\partial u}{\partial T}\right)_v$$

Coefficients

Volume expansion $\beta = \dfrac{1}{v}\left(\dfrac{\partial v}{\partial T}\right)_p$

153

Thermodynamics and Fluid Mechanics

Compressibility $\quad \kappa = -\dfrac{1}{v}\left(\dfrac{\partial v}{\partial p}\right)_T$

Joule–Thomson $\quad \mu = \left(\dfrac{\partial T}{\partial p}\right)_h$

$$c_p - c_v = \beta^2 \, Tv/\kappa$$

For a perfect gas, $c_p - c_v = R$ and $\mu = 0$.

Equations of state

Perfect gas

$$pv = RT = R_0 T/M$$

for a gas of molecular weight M with gas constant R.

Van der Waal's gas

$$(p + a/v^2)(v - b) = RT$$

where a and b are constants of the gas.

Reversible polytropic process

$$pv^n = \text{constant}, \qquad w = \frac{p_2 v_2 - p_1 v_1}{1 - n} \quad (n \neq 1)$$

For a perfect gas,

$$w = \frac{R}{1 - n}(T_2 - T_1), \qquad q = \left(c_v + \frac{R}{1 - n}\right)(T_2 - T_1)$$

$$\frac{T_2}{T_1} = \left(\frac{p_2}{p_1}\right)^{(n-1)/n}$$

here, $n = \gamma$ if the process is adiabatic, i.e. if $q = 0$.

Reversible isothermal process

$$q = T(s_2 - s_1)$$

for a perfect gas, $pv = \text{constant}$ and

$$q = w = RT \ln(v_2/v_1)$$

Velocity of sound in a perfect gas

$$c = (\gamma p/\rho)^{\frac{1}{2}}$$

154

Thermodynamics and Fluid Mechanics

FORMULAE FOR FLUID MECHANICS

Strain energy

$$E = \tfrac{1}{2}p^2 V/\kappa$$

Viscosity

$$\eta = \tau/c'$$

where τ is the viscous resistance per unit of area parallel to the flow and c' is the velocity gradient perpendicular to the flow.

Kinematic viscosity

$$v = \eta/\rho$$

Pressure at depth h

$$p = g\rho h$$

Flow velocity through a cylindrical tube of radius r and length l

$$c = \pi p r^4/8\eta l$$

Velocity of a surface wave of wavelength λ in an incompressible inviscid fluid of depth h

$$c = (gh)^{\frac{1}{2}} \quad (\lambda \text{ much larger than } h)$$
$$c = (g\lambda/2\pi)^{\frac{1}{2}} \quad (h \text{ large})$$
$$c = (2\pi\sigma/\lambda\rho + g\lambda/2\pi)^{\frac{1}{2}} \quad (\lambda \text{ small})$$

where σ is the surface tension.

MECHANICS

Notation

a	size parameter (length)	t	time
f	acceleration	T	kinetic energy
F	force	u, v	velocity
H	momentum	V	potential energy
m	mass	μ	coefficient of friction
M	moment	ω	angular velocity
r, s	position	$\dot{x}$	time derivative of x

Bold letters indicate vector quantities (which may be represented by scalars when constrained to one dimension).

Thermodynamics and Fluid Mechanics

Laws of Coulomb friction

(i)   The friction force developed is independent of the magnitude of the area of contact.

(ii) The friction force is proportional to the normal force.

(iii) At low velocity of sliding, the friction force is independent of the velocity.

Belt friction

The ratio of tensions at the ends of an arc of belt sustending an angle θ is

$$T_1/T_2 = e^{\mu\theta}$$

where μ is the coefficient of friction.

Newton's laws

(1) Every body stays in a state of uniform motion in a straight line unless it is acted on by a force which may change that state.

(2) The rate of change of momentum with respect to time is equal to the force producing it. The change takes place in the direction of the force.

(3) To every action there is an equal and opposite reaction.

Particle dynamics

Impulse and momentum

$$\int_{t_1}^{t_2} F \, dt = m v_2 - m v_1.$$

Moment of momentum

$$r \times F = \frac{d}{dt}(r \times mv)$$

where $r \times mv = h_0$.

Period and frequency

$$\tau = 2\pi/\omega, \qquad v = \omega/2\pi$$

SHM systems

For a sprung mass in linear vibration,

$$\omega^2 = k/m$$

where k is the spring stiffness.

For torsional vibration, with $x =$ angular displacement,

$$\omega^2 = c/I$$

where c is the torsional stiffness and I is the moment of inertia.

For simple pendulum of length l,

$$\omega^2 = g/l$$

For compound pendulum, suspended about a horizontal axis through OX distant l from the centroid,

$$\omega^2 = mgl/I_{ox}$$

Conservation of momentum

If the only forces acting on a system of particles are mutual interactions, then the momentum of the system is constant.

Work and energy

$$\int_{r_1}^{r_2} F \cdot dr = \tfrac{1}{2}mv_2^2 - \tfrac{1}{2}mv_1^2$$

Potential energy

If $F = \operatorname{grad} \phi$, then

$$\int_{r_1}^{r_2} F \cdot dr = \phi_2 - \phi_1$$

and the change in potential energy is

$$-\int_{r_1}^{r_2} F \cdot dr = V_2 - V_1$$

For an inverse-square law,

$$F = (\alpha m/r^2)e_r, \qquad V = \alpha m/r$$

where α is a constant of the field of force acting on the particle of mass m.

Conservation of energy

For a conservative system, the sum $V + T$ is constant.

Uniform acceleration in a straight line

$$v = u + ft, \quad v^2 = u^2 + 2fs, \quad s = ut + \tfrac{1}{2}ft^2$$

where u is the initial and v is the final velocity.

Uniform motion in a circle

$$f = -\omega^2 r$$

Simple harmonic motion (SHM)

Basic equation

$$\ddot{x} + \omega^2 x = 0, \qquad x(0) = a, \quad \dot{x}(0) = 0$$

The solution is

$$x = a \cos \omega t$$

and the velocity is given by

$$v^2 = \omega^2 (a^2 - x^2)$$

ELECTRICITY

Resonance, Q-factor, and bandwidth of electrical circuits

Series resonant circuit

The resonant frequency (minimum Z, or real Z) is

$$\omega_0 = \frac{1}{\sqrt{(LC)}}$$

and the damped natural frequency is

$$\omega_n = \sqrt{\left(\frac{1}{LC} - \frac{R^2}{4L^2} \right)} = \omega_0 \sqrt{\left(1 - \frac{1}{4Q^2} \right)}$$

where

$$Q = \frac{\omega_0 L}{R}$$

is the magnification or

$$\frac{\text{voltage developed across } L \text{ or } C}{\text{voltage across whole circuit}}$$

Electricity

or

$$\frac{2\pi \text{ (energy stored in } L \text{ or } C}{\text{voltage across whole circuit}}$$

or

$$\frac{2\pi \text{ (energy stored in } L \text{ or } C)}{\text{energy dissipated per cycle}}$$

Impedance just off resonance ($\omega_0 \pm \Delta\omega$) is given by

$$Z = R(1 + j2fQ)$$

where

$$f = \frac{\Delta\omega}{\omega_0} \ll 1$$

If $2fQ = \pm 1$, then $|Z| = \sqrt{2}R$ and, for constant voltage, the current is $1/\sqrt{2}$ of its maximum value and the power is halved. These points at which $2fQ = \pm 1$ are known as half-power points and the total bandwidth between them is $2\Delta\omega$. Hence

$$\text{half-power bandwidth} = \frac{2\Delta\omega}{\omega_0} = 2f = \frac{1}{Q}$$

Parallel resonant circuit

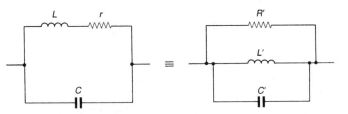

These circuits are equivalent with $L' = L$, $C' = C$, and $R' = Q^2 r$ if $Q \gg 1$. The right-hand version gives results similar to those for the series resonant circuit.

Q becomes $Q' = R'/\omega_0 L'$ and represents the current magnification, and the admittance is

$$Y' = \frac{1}{R'}(1 + j2fQ')$$

159

BLACK-BODY RADIATION

The power radiated by a black body in all directions, per unit surface area and per unit frequency interval, is

$$E_v = \frac{2\pi h v^3 / c^2}{e^{hv/kT} - 1}$$

in the region of the frequency v, where h is Planck's constant, k Boltzmann's constant, and c the velocity of light; alternatively, the power per unit wavelength interval in the region of the wavelength λ is

$$E_\lambda = \frac{2\pi h c^2 / \lambda^5}{e^{hc/\lambda kT} - 1}$$

The total power per unit area is

$$E = \int_0^\infty E_v \, dv = \int_0^\infty E_\lambda \, d\lambda = \sigma T^4$$

in which σ is the Stefan–Boltzmann constant. The wavelength λ_m at which E_λ is a maximum is given by the *Wien displacement law*:

$$\lambda_m T = 0.0029$$